THE GENERAL PRINCIPLE

OF RELATIVITY:

IN ITS PHILOSOPHICAL AND HISTORICAL ASPECT

BY

H. WILDON CARR

©1920

THE GENERAL PRINCIPLE OF
RELATIVITY

MACMILLAN AND CO., Limited
LONDON · BOMBAY · CALCUTTA · MADRAS
MELBOURNE

THE MACMILLAN COMPANY
NEW YORK · BOSTON · CHICAGO
DALLAS · SAN FRANCISCO

THE MACMILLAN CO. OF CANADA, Ltd.
TORONTO

COPYRIGHT

GLASGOW : PRINTED AT THE UNIVERSITY PRESS
BY ROBERT MACLEHOSE AND CO. LTD

PREFACE

It may seem a bold undertaking on the part of one who can claim no acquaintance with the higher mathematics and no familiarity with the experimental work of the physical laboratory, to propose to interpret a principle which is the practical concern only of the mathematician and physicist. It needs no apology, however, for though the principle of relativity has been formulated by mathematicians and physicists purely as a working principle in mathematics and physics, the particular concepts with which it deals—space, time and movement—are metaphysical, and the essential concern of philosophy.

In this account of the principle of relativity I have dealt only with the philosophical and historical aspect of the problem. I have tried to expound the reformed concepts of space and time and movement which are the justification and the foundation of the new working formulae.

I have not attempted to indicate or explain, even in non-mathematical terms, the formulae themselves. I have not, for example, tried to show how Einstein worked out the formula of the precession of the perihelion of Mercury, the displacement of light from stars observed in the eclipse observation, or the shift of the spectral lines. What I have tried to show is the exact meaning in philosophy of the new concept of the framework of nature.

My interest in the principle of relativity is purely philosophical, but it is not casual or accidental. I first became acquainted with it at the International Congress of Philosophy at Bologna in 1911, when M. Pierre Langevin, Professor of the Collège de France, revealed its philosophical importance in a remarkable paper entitled " L'évolution de l'espace et du temps." I introduced the subject to the Aristotelian Society in a paper read in the Session of 1913-14 (*Proceedings of the Aristotelian Society*, Vol. XIV.), and I contributed an article, " The Metaphysical Implications of the Principle of Relativity," to the *Philosophical Review* of January 1915. Since then the philosophical importance of the principle has received full recognition. It was not, however, until the preparation of my courses of lectures on the "History of Modern Philosophy" delivered in 1918 and 1919 at King's College, London, led me to read anew the works of Descartes and Leibniz that the quite special historical interest of the problem impressed me. It is this historical aspect of the principle to which I have tried to give expression in this study. The main idea was developed in a course of lectures on "Historical Theories of Space, Time and Movement" delivered at King's College in the spring of this year (1920).

My thanks are due to Professor T. P. Nunn and Dr. C. D. Broad, who have rendered me special service in reading my proofs. They are not of course responsible for my views or for the accuracy of any of my statements.

CONTENTS

CHAPTER VII

CHAPTER VIII

CHAPTER I

SPACE, TIME AND MOVEMENT

THE new theory of Einstein which is known as
the general principle of relativity is perfectly
simple when once it is understood and peculiarly
difficult to understand. This arises from the
fact that the human mind, in its ordinary attitude
of reflection, and particularly in its well-balanced
moods, subject to reason and superior to emotion,
is always ready to revise its conclusions. When,
however, it is required not merely to revise
its conclusions but actually to amend its
premises, a kind of mental giddiness is experienced, a feeling of insecurity as though the
firm ground on which its conclusions are based
and from which they derive their whole strength
had begun to shake and prove unstable. The
wonderful structure of physical science, with the
assurance consequent on the continual progress
and constant acceleration of its advance in the

A

last two centuries of the modern period, seems in jeopardy the moment real doubt is thrown on the concepts of absolute space and time and movement, which appear as its conditions. It is because these concepts are rejected by the new principle that the revolution in science is so profound and far-reaching.

Space, time and movement seem direct self-revealing realities and to the ordinary man the necessity of having theories about them is difficult to appreciate. There are indeed, as everyone knows, puzzling psychological problems and even perplexing philosophical questions concerning them, but these all seem, when we reflect on them, to concern wholly and solely our knowledge, and the mistakes and illusions which may arise in regard to our knowledge. As to the realities themselves they present themselves as the simple and obvious framework of the objective world of our daily experience and as the subject-matter of mathematical and physical science. We may know perfectly well that many philosophers, following Kant, have held that space and time are forms of perception which the mind possesses as pure *a priori* cognitions. But then this is a theory of knowledge, and the

conclusion which Kant drew from it, that knowledge is of phenomena and not of things in
themselves, leaves the whole reality of physical
science unaffected. We may know, too, that
some philosophers have denied the reality of
movement, while others have denied the reality
of everything which is not movement. But
such opinions are dismissed by us as logical
problems which concern meanings and which
leave the facts of experience unaltered. It is
therefore with considerable perplexity and with
unfeigned surprise that the scientific world
has received the evidence put forward, not by a
speculative philosopher but by a mathematician
and physicist, that our ordinary accepted notions
of space, time and movement do not correspond
with reality, and that the laws of nature require
to be all reformulated on a new principle which
rests primarily on the rejection of space and time
as constant factors.

To the metaphysician there is nothing subversive or revolutionary in the new principle, it
is practically identical with principles which
have, time and again, been formulated in philosophy, ancient and modern, but to the man of
science it seems like a sudden upheaval of the

foundations on which the whole stupendous structure of modern science has been reared.

Einstein's principle of relativity has two distinct stages. The first formulation, in 1905, expressed the acceptance of the consistently negative results of experiments contrived to determine absolute velocity by reference to a fixed system at rest, such as the ether of space was generally supposed to be. If there is no zero system with reference to which absolute velocity can be measured, we have to correlate observations for systems moving relatively to one another. The special principle of relativity, or the restricted theory, is so called because it applied only to uniform rectilinear translations of reference-systems, and not to rotations or non-uniform translations. The special principle is that the velocity of the propagation of light *in vacuo* is constant for every observer, that it is unaffected by the translation of a reference-system relatively to other systems, and that the constancy of the velocity is maintained by a variation of space and time. In 1917 Einstein formulated the new principle of generalized relativity. This was the extension of the earlier principle to include the law of gravitation and by implication all laws of nature.

It accepted for all the laws of nature the impossibility of any absolute standard of reference, and it proposed to determine all universal laws as observational facts to be deduced from the movements of various systems relatively to one another. It involved the rejection of Newton's concept of the attraction of masses acting at a distance from one another in a uniform space and even flowing time, and the denial that any spatial or temporal dimensions are uniform and absolute for all systems of movement. It also rejected the postulates of Euclid as impracticable.

To make the full significance of this new principle appear and to show its philosophical importance in the world-view it discloses to us, is the aim of the present historical study. It will be sufficient, before trying to follow the problem from its origin, to indicate clearly the two facts which the special and the general principle take to be conclusively established. They are both negative facts, and therefore have none of the simplicity of new discovery of the hitherto unknown. They do not give us new notions but they upset our old notions and complicate and render difficult the necessary reconstruction of the world-view.

The first fact is that the velocity of light is a finite velocity, and yet absolutely uniform for every observer, whatever the velocity of his system, and whatever the direction of its movement of translation relatively to other systems. The nature of light is not in question, for whether we accept the corpuscular or the undulatory theory, we know that light is propagated in a movement radiating outwards in every direction from its source, which is thus always the centre of a sphere. The velocity of the propagation of light in empty space was discovered in the early part of the seventeenth century, when the telescope revealed the moons of Jupiter and enabled calculations to be made by comparing the time table of the satellites for the planet when at its nearest point and when at its most distant. The interval of time and the distance traversed, both being known, gave the velocity of light. It is a velocity which for all terrestrial distances is negligible, it only becomes of account in the great spatial intervals which separate planetary and stellar masses. We have no other means than that of light signals to enable us to determine the simultaneity of events, and yet light signals are themselves subject to the interval required for their transmission between the observers at distant points who are using them. Now, if space and time are absolute, as we ordinarily suppose, then when distances and intervals are varying by reason of the movements of the observers relatively to one another, it is quite clear and evident that the velocity of light for the observers must vary correspondingly. But experiments specially designed for the purpose have proved conclusively that the velocity of the propagation of light does not vary, it is uniform for all observers whatever the relative movement of the systems in which they are situated. Let us take an extreme case and suppose that two observers of the same events are in different systems of reference, and that each observer, thinking himself at rest, sees the other system moving with a translation of 100,000 miles a second, that is, rather more than half the velocity of light in empty space. Now it would be rational to conclude, and we should naturally expect to find, that if these two observers communicated with one another by light signals, the velocity of the propagation of the light signal would be more than twice as great, in the direction of the uniform movement, for the one observer as it would be for the other. This is found not to accord with experimental fact. So the principle of relativity declares that the velocity of light is constant, however the conditions of the observer vary by reason of the translation of the system, and that space and time are different for different systems. To ordinary reason this is a paradox. Einstein has accepted the experimental proof without any attempt to explain it away as appearance or illusion. He formulated the principle of relativity to accord with the result of the experiments. The principle is then, that the velocity of light is constant and that space and time are variable. I am not at present inviting attention to, or challenging criticism of, the evidence for this fact, so subversive of ordinary ideas and upsetting to our habits, I am trying only to state as definitely as possible what the fact is. Certainly in the case of the enormous velocity of light and the infinitesimal fraction of it represented by any known velocity of translation, the fact, if we accept it, is negligible as applied to our common terrestrial life, but it is very difficult indeed to reconcile with our experience of velocity generally. Sound, for example, is a propagated movement, but when the source of sound is moving with us, as when we are talking in an open motor-car, we naturally adapt ourselves to the idea that the sound waves are not spreading with equal velocity forward and backward. We think, wrongly perhaps, that the waves of sound, when we in the car are moving in their direction, spread out from the car at a lower velocity than when we are not moving with them. The special or restricted principle of relativity then is, that the velocity of light is constant for all observers and independent of their system of reference, and that space and time are variable, dependent on the relative translation of systems.

The general theory of relativity goes much further. It extends the principle to all the laws of nature. It rests upon a fact or rather upon a negative discovery,—a discovery which is not due as in the case of special relativity to definite test experiments but the result of the successful application of the principle to the formulation of a new law of gravitation. The proof of the new principle rests on the fact that it has been found to account for a well-known discordance between the astronomical calculation for the precession of the perihelion of Mercury and the actual observation, which had previously baffled all attempts to explain. Further, it enabled a prediction to be made as to the deflection of the light from a star passing near the sun during a total eclipse, which was verified in the observations of the eclipse of the sun in May 1919. A further prediction by Einstein that the spectroscopic analysis of atoms vibrating in the gravitational field of the sun compared with the analysis of similar atoms on the earth would show a shifting of the lines towards the red end of the spectrum has at present not been verified and is the subject of research. It is not, however, with the details of these tests of the principle which I am now concerned. I want rather to make plain the fact which is alleged as the basis of the new theory. As applied to the new theory of gravitation it is called equivalence.

If we raise an object and then release it, it drops. We explain this as an instance of a law of gravitation by which bodies attract one another in a definite relation of their mass and distance. We regard the floor as fixed in relation to the earth, and the released object falls to it, drawn, we say, by the attraction of the earth. But the earth, to which the floor and the room are attached, is rotating on its axis; it is also travelling on its orbit at many miles a second; and the whole solar system is moving in the stellar system. It is clear, therefore, that there might be an observer who would say that the released object remained at rest and that the floor of the room moved to it. The theory of equivalence is that there is no way of deciding between the alternative descriptions, whether in fact the object fell to the floor or the floor rose to the object. If one observer had the right to decide positively for the one, another observer would have the equal right to decide positively for the other. If the principle be accepted, it completely negatives the idea that forces of attraction are exercised by bodies on one another in the sense supposed in Newton's law.

If this negative fact be established, namely, that there is no way of determining the actual line which two objects follow in their movement towards one another, and that contradictory descriptions of such movement are really equivalent, it follows that space cannot have the properties which Euclid required, and force cannot have the nature which Newton supposed. The discovery can only be compared in importance with the discovery of Copernicus that the earth is not at rest but undergoing a diurnal rotation on its axis and an annual revolution round the sun.

Space and time are concepts of the mind. They are indeed for all of us realities with which we feel we are in direct relation, a relation so fundamental that our whole existence depends on it. But space and time in themselves, though not abstractions, lack the concreteness of objects and events. They are a framework of the physical universe and give form and continuity to its content. As concepts they are judged by their consistency or inconsistency. The dominant place they occupy in philosophy, and the persistence of the problems they give rise to throughout the whole history of philosophy, ancient and modern, are due to the inherent logical and metaphysical difficulties they present. But space and time are not only concepts, they are also images. In studying the theories of space and time it is very important to take into account the imagery which supplies to the concepts their content. It is usual to neglect this completely. The reason is that philosophers never reveal the imagery which lies behind and supports the concepts they analyse. For imagery we have to go to the poets. *Pari passu* with the evolution of the concepts of space and time has been an evolution of their images. Homer, Dante and Milton are as distinct in the imagery of their expression of a world-view as Aristotle, Aquinas and Leibniz are in their concepts of its reality. Every philosopher starts his reflection from the stand-point of his world-view. The world-view is an imaginative background of his thoughts, his reflections borrow their shape and draw their content from it, revolve round it and always return to re-form it. But when we study a philosopher's theories we treat them in the mathematical method, substituting signs for images. We suppose there is a special advantage in this power of detaching the sign completely from the image in which it arose. As soon as we grasp a man's concept we adapt it to our own imagery, whatever it may be, and proceed as though the world-view were of no importance. A familiar illustration is the way in which the Bible is interpreted in Christian households. The concepts are detached from the imagery of the writers and fitted on to the homely imagery of the reader whatever it may be. We study in the philosophers their logical principles and abstract concepts, in the poets their imagery, we forget that the poets express the imagery which the philosophers require to embody their concepts.   If we would reconstitute the thought of an historical period we must read its poetry in conjunction with its philosophy. When we discuss to-day the theories of Newton, we take no account of the world-view which presented itself to him and of its complete difference from our ordinary world-view to-day.   Our world-view is continually changing, and the imagery in which we clothe it becomes outworn and cast aside.   How completely different, for example, is the world picture presented to us in Mr. Wells's *Outlines of History* from anything which filled the imagination of a previous generation.   I have chosen Newton as an illustration because we are accustomed to accept his concepts as essentially modern.   Science has advanced, but his concepts remain of universal application. Newton's age is so near our own, as compared with the Greek and Mediaeval ages, that we hardly appreciate how much its imagery has changed.   Yet how fantastic the world-scheme of Milton's *Paradise Lost* appears to us to-day and how inadequate his imagery to embody

modern concepts. It was, however, the familiar
background of Newton's thoughts.

> " Now had the Almighty Father from above,
> From the pure empyrean where he sits
> High throned above all height, bent down his eye,
> His own works and their works at once to view.
>
> .   .   .   .   .   .
>
> On Earth he first beheld
> Our two first parents, yet the only two
> Of mankind, in the happy garden placed,
> Reaping immortal fruits of joy and love.
>
> .   .   .   .   .   .
>
> He then surveyed
> Hell and the gulf between, and Satan there
> Coasting the wall of heaven on this side night
> In the dun air sublime, and ready now
> To stoop with wearied wings, and willing feet,
> On the bare outside of this world, that seemed
> Firm land imbosomed without firmament,
> Uncertain which, in ocean or in air."

Unlike Dante's world, heaven and hell have
no direct connexion with our universe, which is
conceived as a system of sun and planets swinging
in vast space, yet an ordered system with laws of
nature imposed upon it. It is a new creation,
espied from afar by Satan, and offering, in its
order and arrangement, rest for wearied wings
and a sphere for concerted action. But what
strikes us particularly in such imagery, as

compared with that which we should now deem adequate, is that the distant observer surveying our world sees it as it appears to us and makes no allowance for systems of reference. Our spatial and temporal coordinates are also those of God and of Satan. This was essentially Newton's view.

The importance of imagery and the way in which it qualifies concepts may be illustrated also in a somewhat different way. Take the case of the familiar phenomenon of the ebb and flow of the tide which we explain by the concept of gravitation. For us the tides mean an alternate rise and streaming of the water in one direction and a fall and streaming in the reverse direction, with all its minute and dependent circumstances. To an outside observer the tide would mean only the unalterable shape a plastic body in rotation would assume in spite of the changing position of the mass.

Throughout the whole history of human thought, while imagery and concepts have been changing continuously, the fundamental notions of space and time and movement, both as being direct data of experience, and necessary conditions of experience, have withstood all change. They appear to us as the framework of our universe, whatever the content and the nature and the history of that universe be. And yet from the very beginning of our historical records of human reflective thought, everyone who has turned his thoughts upon them has found that they present insoluble problems and offer the strangest paradoxes.

Neither our images nor our concepts of space and time are identical with anything spatial or temporal which we perceive. It is from this incongruence of percepts and concepts of space and time that the psychological problems in regard to them arise. Space is imaged either by its negative character as the void or by its positive character as extension. But neither void nor extension is direct experience or a datum of sense-intuition. Although space and time are intimately bound up with all sense experience, there is no actual sense experience of space and of time. We cannot, for example, satisfy in regard to the ideas of them a demand such as Hume proposed for a universal test, produce the impression which has given rise to the idea. Of space and of time there are no impressions. A still more surprising and even disconcerting fact is that while image and concept of space and time fulfil completely the Euclidean postulates and conform exactly to the axioms, not one of our senses gives us spatial and temporal experience conformable to those conditions. In his *New Theory of Vision* Berkeley proved that the sense of sight cannot yield a perception of distance or give us knowledge of the third dimension of space, and based on this the theory that visual perceptions are a language of signs, the purpose of which is to enable us to anticipate tactile sensation. But tactile sensation will not, any more than will visual sensation, give us knowledge of distance, such knowledge depends on movement, and movement involves time as well as space. If so, then what is the absolute standard by which we are to measure time? Try in what way we will, we can never by direct perception arrive at the notions of absolute space and time which yet we imagine and conceive to be the basis of the reality of nature.

This is no new discovery. It is indeed a commonplace of philosophy and even of the modern science of psychology. One of the large problems in contemporary psychology concerns the nature and origin of the perception of

space. There are numerous theories, which fall
however into two main groups. They are named
the genetic and the nativistic theories. The genetic
theories derive our notion of space from sense
experience which is not itself spatial, by means
of inference and mental construction. The
nativistic theories, on the other hand, derive it
from the mind itself and the mode of its activity
in experience.

A genetic theory has been held by most of the
older, as well as by many of the present, generation
of modern psychologists. An illustration of it
is the theory expounded by Herbert Spencer
(*Principles of Psychology*, ii. 178), according to
which the perception of space is simply an
interpretation of the simultaneity of sensations,
explained physiologically in the case of sight
by the overlapping of successive stimuli on the
retina and in the case of touch by the reversibility
of series of tactile impressions. Another illustration is the well-known local-sign theory of Lotze.
The local sign is not a localization or extension
in the sensation itself, but a character belonging
to tactile impressions which later causes the
mind to locate them in particular points of the
body. It is from these impressions that our

mind is supposed by the theory to construct the perception of space.

An example of nativistic theory is the view expounded by William James in *Principles of Psychology* (vol. ii. p. 134 ff.) that there are sensations to which the character of voluminousness distinctly belongs and which are thereby able to give the mind direct perception of space. This character, called by other psychologists extensity, is not extension, that term being only applicable to physical objects. Extension is a sensible quality, extensity is a character of sensations.

It is not then in philosophy nor in the science of psychology that the principle of relativity is revolutionary. It is only a revolution in physical science, and it is a complete revolution in science, because mathematics and physics have seemed justified in rejecting, as outside their sphere and completely indifferent to them, the problem of the relation of the mind to its objects. The objective character of physical science, upon which it has prided itself, has therefore come to mean the uncritical assumption of absolute space and time. The introduction into pure mathematics or into pure physics of a subjective element seems not only a sacrilege but a downright betrayal of the very principle on which science is based. It has been supposed that in its purely objective basis lies the strength of physical science and that to this objective basis is due the steady and rapid and continuous progress which is often vaunted as presenting a favourable contrast to speculative philosophy.

When the principle of relativity was first formulated it was generally put forward as a methodological principle applicable only within the sciences concerned and with no relation whatever to any question of general philosophical or metaphysical theory. It simply, it was said, proposed a reform of mathematical procedure, a reform which was radical indeed, for it involved, not the correction or improvement of the accepted equations, but a new set of equations involving new constants and new variables. The general principle of relativity now proposed by Einstein is acknowledged, however, to concern the most fundamental philosophical concepts of the nature of the universe. The essence of it is to introduce the bane of the physicist, subjectivism, into the arcana of physical science. It shows that it is impossible to abstract from the mind of the observer and treat his observations as themselves absolute and independent in their objectivity. It requires us to give up the assumption of an absolute standard of reference for the measurement of the velocity of a system. It rejects the inference, which all our experience and all our science has seemed with such increasing assurance to affirm, that beneath the objects we perceive, juxtaposed in the external world, there is an absolute space which would be void, but not abolished, if they were removed, and that behind the events which succeed one another in our consciousness, there is an absolute time which might lose all distinction if there were no events, but which would still flow. We are to reject this inference not because it is found to be useless, not because pure space and pure time are undiscoverable, not because we can never by direct perceptive means become acquainted with them, but because physical experiments which ought to have revealed them if they exist, have uniformly failed to do so. The new principle is not a belated discovery of our ignorance ; it is a new advance in positive knowledge. In this lies its strength. The study of nature has revealed to us that the nature we study is not independent of the mind which studies it. There is no absolute physical reality which a mind may contemplate in its pure independence of the contemplator and the conditions of his contemplation. The new principle is that every observer is himself the absolute, and not, as has been hitherto supposed, the relative, centre of the universe. There is no universe common to all observers and private to none. The work of physical science is to co-ordinate the observations of observers, each of whom uses his own co-ordinates and for whom there is no common measure.

CHAPTER II

THE ANTINOMY OF MOVEMENT

ARISTOTLE in the *Physics* (vi. 14) says that Zeno committed a fallacy when he argued : " If everything in order to be, must, whether moving or at rest, occupy an equal space, and if a body when displaced occupies at every moment an equal space, then it follows that the flying arrow is immobile." It is an error, Aristotle argues, because time is not composed of moments, that is, of indivisibles. Neither indeed, he adds, is any other magnitude.

Whether or not Aristotle's refutation of Zeno's argument is sound, it is certain that philosophy generally has not found it possible to dismiss the problem of movement in this summary way. Many philosophers indeed have been equally confident, but a glance at the history of philosophy shows the problem cropping up in some form in every stage of the evolution of the concept of metaphysical reality.

Zeno's famous arguments against movement are four in number, and together they are so compact that those who would refute them look in vain for a logical loophole. The first declares that it is impossible that a body can move from one point to another distant from it, because, before it can traverse the whole intervening space it must pass through half, and before it can traverse that half, the half of the half, and so on, to infinity. The second is that Achilles in his race with the tortoise can never overtake it, if it is allowed to have a start, for to do so he must first reach the point at which the tortoise is, but when he reaches it the tortoise will have moved on, and Achilles, therefore, will have always a step to take. The third is that the flying arrow does not move because at every moment it is at rest. The fourth is that if there are three processions in the stadium, each composed of equal numbers and equal masses, one of which remains stationary while the other two move with an equal velocity but in an opposite parallel direction, passing the first in mid course, then it follows that each moving procession will traverse an identical space in a time which will be both half and double of itself.

The last of these arguments can be made quite clear in a diagram. Let us suppose $A_1A_2A_3A_4$, $B_1B_2B_3B_4$ and $C_1C_2C_3C_4$ to be the three processions. Let us suppose their first position to be

$$A_1A_2A_3A_4$$
$$B_4B_3B_2B_1$$
$$C_1C_2C_3C_4$$

The A's are stationary, the B's are moving to the right, the C's to the left. When then B_1 reaches A_4, C_1 will reach A_1, and their position will be

$$A_1A_2A_3A_4$$
$$B_4B_3B_2B_1$$
$$C_1C_2C_3C_4$$

But in reaching this position the C's will have been consecutively in line with all the B's and with half the A's, and the B's will likewise have been in line with all the C's and with half the A's. But B's and C's and A's occupy equal spatial magnitudes. The difference therefore is not in the space. The time also is identical for it is one and the same interval, yet it is only half for the B's and C's what it is for the A's ;

the half, therefore, is identical to the whole, or
the time is the double of itself.

The argument may be put in another form
which perhaps is even more perplexing. Suppose
the processions to be points and the succession
instants, that is, suppose the divisions of the
movement to be units of time and space. Suppose
the position at a first instant to be

$$A_1A_2A_3A_4$$
$$B_4B_3B_2B_1$$
$$C_1C_2C_3C_4$$

and at the second instant (when the B's have
moved one point to the right, the C's one point
to the left)

$$A_1A_2A_3A_4$$
$$B_4B_3B_2B_1$$
$$C_1C_2C_3C_4$$

Then at the first instant C_4 is in line with B_1 ;
at the second instant it is in line with B_3, but
it must have passed B_2, and there is no instant
in which it could have been in line with B_2. Also
B_4 is at one instant in line with C_1 and at the next
with C_3—but C_2 lies between, when was B_4 in
line with C_2 ?

Aristotle's refutation of this fourth argument

is of particular interest. " The fallacy consists in supposing that the equal magnitude, possessing the same velocity, moves in the same time both relatively to a mass in movement and relatively to a mass at rest ; therein lies the error " (*Physics*, vi. 14, § 10). By this he appears to mean that while mass and velocity of a moving body remain constant, the time it takes to pass a body at rest and a similar body in movement is not the same. This might be interpreted as an anticipation of the principle of relativity so far as time is concerned, but clearly the very opposite is intended. Aristotle means that time is absolute and that less of it is occupied in passing a mass at rest than in passing an equal mass moving parallel and opposite to it. This, however, leaves Zeno's argument unanswered, merely affirming what Zeno supposes to be affirmed. Zeno says in effect that if movement is real and a body passes from point to point, from moment to moment, then you are committed to the contradictory and absurd assertion that the same time is different.

Zeno lived in the fifth century before Christ, the century which preceded the great philosophical enlightenment represented by Socrates, Plato and Aristotle. He was a pupil of

Parmenides, the head of the famous Eleatic school of philosophy. The rival Ionic school had as its founder Heracleitus of Ephesus. The two schools represented opposite and contradictory principles. According to Heracleitus " becoming," according to Parmenides " being," is the first principle of existence. There is a curious outward resemblance between these early speculations and those of modern transcendental philosophers. The resemblance is in the concepts, and it is a striking illustration of the way concepts abide identical throughout all change of imagery. Moreover, first principles present themselves to reflection as essentially simple and extremely general. It was, however, in their successors that the doctrines of the great founders developed into paradox. Thus the doctrine that all things flow, that reality is universal becoming, was developed into complete paradox by Cratylus, as related by Aristotle in the following description of the Heracleiteans. " And again they held these views because they saw that all this world of nature is in movement, and about that which changes no true statement can be made ; at least, regarding that which everywhere in every respect is changing, nothing could truly be affirmed. It was this belief which blossomed into the most extreme of the views above mentioned, that of the professed Heracleiteans, such as was held by Cratylus, who finally did not think it right to say anything but only moved his finger, and criticized Heracleitus for saying that it is impossible to step twice into the same river ; for he thought he could not do it even once " (*Metaph.* iv. 5). It was this doctrine which Zeno combated.

No one will understand Zeno's arguments who regards him as merely a skilful dialectician and ignores the essential fact that he had reached independently the conclusion that movement is not reality but appearance and used the arguments to enforce it. The arguments therefore are not sophisms nor exercises in logomachy. If you seek his own solution of his paradoxes, it is quite simple. He held that nothing moves, that reality is one and unchangeable.

It should be noticed that the four arguments are cumulative in force. The first shows movement to be impossible, the second shows it to be unreal, the third, contradictory, and the fourth, absurd. The first deals only with space, and the infinite divisibility of space is made the obstacle of movement. The second shows that if movement be supposed actually in progress, contradiction breaks out in the concept of velocity. In the third, discrete points in space are correlated with discrete instants of time, and the contradiction lies in the attempt to correlate the passage from one point to the next with the passage from one instant to the next. It involves the paradox that the arrow is somewhere at no time or nowhere at some time. The fourth combines all the other three, for it takes into account the space, the time and the movement, and it shows that measured by points and instants velocities are infinitely different and all equal.

This was Zeno's problem. It is a problem, therefore, which has its origin in the early Greek nature speculations in which the development of Western philosophy takes its rise, and it is a problem which has persisted throughout the whole of that development and is an unsolved problem to-day. The form, however, has changed. It is as an antinomy of reason that it presents itself to us. No one to-day, even if he argues, as Mr. F. H. Bradley does, that movement is appearance and not reality, is content with the simple denial of movement and the affirmation of the un-changeable one. The antinomy in the concept of movement consists in the fact that the thesis which affirms it, and the antithesis which denies it, present themselves to the mind as equally valid ; yet they are mutually self-contradictory. The thesis is : There are movements, for reality, the reality of life in particular, denotes activity ; a thing is what it does. The antithesis is : There are no movements, for a condition of movement is that a thing which moves shall endure unchanged throughout the movement ; but if nothing changes nothing moves.

The antinomies of reason were made by Kant the central point of interest in the modern philo-sophical problem, so far as it concerns the basis of physical science. According to Kant's theory antinomies arise when the mind makes an object of the whole series of conditions which constitute the system of the world. It is the nature of the mind to present to itself such an object, but the world so presented is an object of reason, not an object of sense intuition nor of understanding. The object of reason is an idea of the unconditioned, it transcends any possible experience and as thing in itself is unknowable. The objects of reason give rise to Ideas (the soul, the world, and God), which have an important function in theory of knowledge, but they are not objects of which we can possess any empirical knowledge. Our interest in them, and their value to us, is practical not speculative. We only know phenomena, not things in themselves.

The antinomies of Kant give us, then, in modern form, the contradictions which lie concealed, or which if known are consciously ignored, in our ordinary common-sense concepts of space, time and movement. Two of the four antinomies, which Kant distinguished as mathematical from the other two as dynamical, are directly concerned with these concepts. The first deals with the self-contradiction involved in thinking of the world as limited or as unlimited in space and time. The thesis is : The world has a beginning in time, and is also limited in regard to space. And the antithesis is : The world has no beginning and no limits in space, but is in relation both to time and space infinite. This antinomy expresses a difficulty which occurs to everyone in moments of reflection. It is impossible to think that the world had no first moment, for in that case how are we to represent the actuality of the present moment, for that moment is a now which ends a series, and its reality therefore seems to depend on a now which began the series ? But then, how on the other hand can we present to the mind a moment to which there is only an after and not a before ? Similarly in regard to space. There is a point " here " which has definite relations to the whole extended universe. The reality of these relations limits the universe. Yet how can we think limits to the universe without in the very thought supposing an extension outside the limits ?

There are two contemporary philosophers, Mr. Bertrand Russell and M. Bergson, who have analysed Zeno's arguments in their original simplicity as the denial of the reality of movement. Mr. Bertrand Russell (*Principles of Mathematics*, chap. xlii., and *Our Knowledge of the External World*, chap. v.) holds that Zeno is right, but that the paradoxical character of the arguments entirely disappears when they are expressed in terms of the modern mathematical theory of infinity. M. Bergson (*Creative Evolution*, pp. 325-330 and *Time and Freewill*, chap. ii.) holds that Zeno's conclusion is wrong in so far as it denies the reality of movement, and that his paradox is due to confusion between a reality in its essential nature indivisible and the intellectual device of a scheme, created and contrived for the practical purpose of division and articulation. The two modes of analysing the old argument and the antithetical conclusions they reach reveal that two principles are contending in philosophy to-day, recalling in a striking way the principles which divided the ancient world, the principle of the unchangeable one and the principle of the universal flow.

Mr. Russell maintains that the paradox is completely solved by the philosophical theory of mathematical continuity. According to this theory space and time actually consist of discrete points and instants, but in any finite portion of space and interval of time the number of points and instants is infinite. In an infinite series no two members are next one another, for between any two there is always another. When accordingly space is conceived as infinitely divisible, this means that the series of points is compact, there is no interstice between one and another. Yet, though there is nothing between the points but points, the points are not next one another, there is no next point to any point. The infinite divisibility of time implies the same of the instants. Having defined continuity in this way it is claimed that all the supposed contradictions in a continuum composed of elements are completely swept away and the foundation laid bare of a reality on which a firm constructive philosophy can be built. The answer then to Zeno is as follows. Zeno asks how can you go from one position at one moment to the next position at the next moment without in the transition being at no position at no moment ? The answer is that there is no next position to any position, no next moment to any moment because between any two there is always another. If there were infinitesimals movement would be impossible, but there are none. Zeno therefore is right in saying that the arrow is at rest at every moment of its flight, wrong in inferring that therefore it does not move, for there is a one-one correspondence in a movement between the infinite series of positions and the infinite series of instants. According to this doctrine then it is possible to affirm the reality of space, time and movement, and yet avoid the paradox in Zeno's arguments.

Bergson's way of escape from the paradox is

entirely different, for it rests on a metaphysical concept of life and a philosophical theory of the nature of the intellect. It does not depend on the mathematical definition of continuity, for mathematical continuity has no relevance to the problem. I mean that as Bergson presents the problem it is indifferent how we describe, or in what terms we define, the continuity of space and time, because it is space and time themselves which are wrongly apprehended. They belong essentially to the intellectual view of reality, while movement as true duration or change is the fundamental reality of life. Take the points and instants of space and time as the elements composing the movement and you will be forced to the conclusion that there is no movement, for the elements are immobilities and movement cannot be generated out of immobilities. But there are real movements, and the immobilities into which we seem able to decompose them are not constituents of the movement, they are views of it.

There are thus two solutions of the antinomy offered to us in contemporary philosophy. I have not included Mr. Bradley's argument in *Appearance and Reality* because it can hardly be classed as a solution. It founds an important philosophical doctrine on the antinomy of movement, but it does so by accepting the contradiction and not by resolving it. There is, however, now offered to us a third and more complete way of escape in the new principle of relativity. This is in effect a reform of the foundational concept of physical reality, and it gives us a new world-view from which the antinomy has disappeared without violence done to reason, or to science, or to common-sense.

If we accept the terms of Zeno's argument there is no escape from the conclusion, and the only salvation from the antinomy lies in successfully attacking the premises. This is what Einstein's theory does. It rejects the concept of absolute space and time. Space and time are not independent of the observer, and there exists no abstract spatio-temporal system by reference to which the velocity, direction and duration of a movement can be absolutely determined. Space and time are variable, and they vary for each observer with his system of reference and with every change in the acceleration of the movement of that system relatively to other systems. Our four-dimensional world preserves its uniformity because our units of length, breadth and depth and our unit of time, alter continually, adapting themselves to the standpoint of an observer at rest, or rather to the standpoint of a system at rest relatively to the translation of other systems.

CHAPTER III

ATOMS AND THE VOID

WE have seen that in the speculations of the early Greeks in nature-philosophy, two opposite and contradictory principles emerged and divided the schools into rival camps. One took " becoming," the other " being " as the first principle of existence. The conflict of these two principles issued in the ancient world in the synthetic construction of a system which has ever since held sway over the human mind. This is the atomic theory of Democritus, of Abdera in Thrace, an older contemporary of Socrates, and the first formulator of philosophical materialism. In so far as the atomic theory is a science of nature there is at every point, despite the enormous advance of physical science in modern times, and the development of means of extending our knowledge by experiment, a most striking consistency between the old atomic theory and the new.

The theory of Democritus is the first attempt of Western thought to present nature as a complete self-contained system. It is a pure materialism for it deduces the whole of the phenomena of the universe, psychical as well as physical, mental as well as bodily, internal and spiritual as well as external and objective, from the concept of an eternal and indestructible matter. There would seem to be a bias towards materialism in the nature of human intelligence, for nothing is able to exorcise completely the hold which it maintains over ordinary experience. Its principle seems eminently rational, and it demands, it would seem, continual and sustained effort to maintain against it what we may have come to regard as stronger reason. Yet although materialism has always appealed to the human intellect as rational and indeed as enforced in some measure by every practical concern of life, it has never held sway for long. Humanity has revolted against it, sometimes with contempt, generally with loathing, too often with passionate hatred. The reason is not that it is irrational, but that it has always seemed to destroy morality at its roots and to sap the foundations of religion. Yet to reject materialism on moral and religious grounds so far from serving philosophy is disastrous to it. If materialism is condemned it ought to be on philosophical grounds alone, and if it be philosophically untenable, everything is unfortunate which tends to conceal its weakness. For my own part, I frankly confess, materialism seems consistent with the highest ethical principles and with the purest religion. I reject it solely on philosophical grounds. Its essential principle not only fails to satisfy me but stands opposed to what appears to me the most obvious truth. Mind is more than matter. In every respect and from every standpoint mind is richer, fuller, greater, more comprehensive. Any principle which proposes to deduce that more from the less stands self-condemned. Yet this is the essential principle of materialism. Given something absolutely self-identical and deprived of difference, materialism declares that by mere external combination and relation there will be produced the variety of the universe including the spiritual values. According to the ancient theory, indivisible atoms, identical in everything but quantity and shape, by their combinations and movements, were held to be able to produce, and in fact had produced, the infinite complexity of the universe. According to the modern theory, simple elements, reducible ultimately to single electrical charges, by mere external combinations in atoms and molecules, are thought to be able to give rise to every form of reality, natural and spiritual.

Our knowledge of Democritus is derived only from references to him in the classical writings, but a very complete account of his atomic theory is enshrined in the great poem, *De Rerum Natura*, of the Roman philosopher-poet Lucretius. In that poem Lucretius has presented to us the philosophy of Epicurus, a philosopher regarded by his followers as divinely inspired and revered as the founder of a religion, or at least of a philosophy practised as a religious duty. Lucretius lived in the first half of the century before Christ, and therefore belongs to the last period of the Roman Republic. Epicurus taught in Athens at the end of the fourth and beginning of the third century B.C. ; Democritus was a century earlier still. The philosophy of Epicurus was an ethical theory. He accepted and adopted the atomic theory of Democritus as the scientific basis of his ethics. Lucretius is a true poet, and the science of nature which he has expounded in his poem is not directly intended for instruction but to support a moral and religious argument. He is moved by a deep love of nature and profound pity for human misery and by a firm belief in the power of philosophy to dispel the principal evils in man's lot. The greatest misery which humanity endures is not physical evil but mental torture due to superstitious fear. Could a man be convinced that the Gods have no interest in human affairs and cannot intervene in the concerns of his earthly life, could he moreover be assured that death is a release and not the beginning of imagined terrors, the two great hindrances to human happiness would be removed. The pleasure which every living creature craves for as part of its nature could at least be enjoyed unspoilt by the poison of superstition. For this purpose he unfolds the philosophy of his almost divine master, and the poem, from the invocation to Venus, not only as goddess of love but as the goddess who has some influence over the cruel God of war, to the close with its terrible description of the plague in Athens, is inspired by a melancholy and deep yearning to alleviate the miserable lot of mankind by an effective deliverance from superstitious fears. The thought that runs through the poem may be gathered from a few examples.

" When we shall have seen that nothing can be produced from nothing, we shall then be able to ascertain correctly what the elements are out of which everything can be produced and the manner in which all things are done without the hand of the gods."

" If things come from nothing, any kind of thing might be born of anything, no seed would be required. Men might rise out of the sea, fish out of the earth, birds out of the sky. Fruits would not be constant to the trees which produce them, any tree might bear any fruit. But instead we see that the rose blooms in spring, the corn ripens in summer, the vintage comes in autumn. If things came from nothing there would be no certain seasons and no time required for growth. Infants would grow at once to men and trees spring in a moment from the ground. But none of these events happen ; all things grow step by step and in growing preserve their kind."

" Moreover nature dissolves everything back into its primitive elements and does not annihilate things."

" If infinite time has not destroyed things it

can only be that things are indestructible."

These passages are from the beginning of the first book, and introduce the theory of the atoms. Another passage may be quoted to show the argument for the void. " The waters, some say, make way for the fish which swim in them, they open liquid paths to them because the fish leave room behind them into which the yielding waters may stream. Thus things may move and change among themselves although the whole seem to be full. But, I ask, how can the fish move forwards unless the waters have first made room ? And on what side can the displaced water go, so long as the fish has not moved ? You must therefore deny motion or admit that the void is mixed up with things in order that motion may get a start."

The science which Lucretius offers us rests on the theory that all things are composed of atoms, that atoms can move by reason of the surrounding void and that all phenomena are produced by the movements of the atoms. It is not exact science as the moderns conceive science, for the ancient philosophers, however acute their observations and precise their descriptions, and ingenious their hypotheses, had neither devised nor developed the experimental method which is the distinguishing feature of modern scientific research.

The ancient atomic theory arose directly from the paradoxes of the rival principles of the old nature-philosophy. It was the great and profound synthetic work of a man of genius. It was worked out into a complete system, and as a perfect expression of materialism it has exerted a continuous influence throughout the whole history and development of Western thought.

We can see how the system of Democritus arose. If all things flow, some simple unchangeable being must support the movement. If this being moves, non-being cannot be nought. Movement is impossible if everything is divisible, therefore if there is movement there is a limit to divisibility. Movement is unreal if the indivisible atoms fill all space, for then movement is blocked ; therefore, besides the atoms there is void. Movement is contradictory if the moving body is at every moment at rest, therefore there is some force which causes the atoms to move. There must be persistence of matter throughout the infinite variety of changing form. The atoms are the identity unchanged beneath the varying wealth of sensory appearance.

The principle of materialism is that the simplest explanation is the best, and the atomic theory deduces all the wealth of existence from absolutely simple beginnings. It is helped by an analogy. Just as the infinite variety of literature is produced by means of the letters of the alphabet, which undergo no change throughout all their combinations, so we may suppose that the phenomena of the universe with their infinite variety of colour and form can be reduced in the last analysis to very simple elements, almost identical, yet able to produce variety in profusion simply by combinations. These simple elements are the atoms ; by uniting and combining they form material objects ; by changing their place they bring about the continual shifting of phenomenal change.

What can we know positively about atoms ? We cannot see them, nor by any conceivable means make them evident to the senses. Not only has no one seen an atom, but we can be certain that no one ever will ; for anything large enough to be an object of sense-perception would not be indivisible. The concept of indivisibility places them far below the limit of perception, and the fact of indivisibility assures us of their reality. Collected into a group of sufficient members they form a body which can be seen and touched. The fact of their existence is thus derived from the necessity of denying the infinite divisibility of being.

The quantity of the atoms is infinite, for there is no limit to the bodies which the universe contains. Also they are eternal and indestructible. This also follows from the concept of them. They have no other quality than their form or shape. In this alone is their difference from one another. Colour, odour, weight, resistance are due to their combinations and movements. The unchangeableness of the atom follows therefore from its nature. It has been and will be what it is throughout eternity. How can it change seeing that it is indivisible ? How can it alter its quality, seeing that it has none ?

Bodies which are composed of the atoms appear to us coloured, resistant, sonorous, hot or cold, but this is illusion, for these qualities are the impressions on our sense organs and therefore appearances. Dissipate the illusion, think of bodies as they must be in

D

themselves, and it will be seen that they must consist of atoms and that atoms cannot themselves possess the sense qualities. But just as atoms have different forms so the bodies composed of them will be different according to the arrangement and direction of the atoms in them. When, then, one identical body appears different at different moments it is because its atoms have changed place or because some of its atoms have been lost or gained. It is analogous to the case of words which alter and change in both sound and meaning by the addition or subtraction of a letter or the alteration of the arrangement of the letters.

Such then are the atoms—how have they come to form the world in which we live? We must suppose that the atom left to itself in the void would have a natural movement, a movement inherent in it, a weight or gravity which would cause it to fall for ever in the infinite void. From this it will come about that from time to time atoms will clash, will block one another and form conglomerations or heaps. Our world must be conceived as such a heap, and by the clashing, sorting, collecting and dispersing of its atoms, there have successively formed themselves, the earth floating in the air, the moon and the sun which are bodies like the earth, the stars, and also the living beings on the earth. The soul which appears to animate the organized bodies must be supposed to consist of more subtle atoms, very mobile, which we may imagine to be round and polished and so comparatively frictionless.

The thoughts which succeed one another in the soul are the movements of its component atoms. Democritus seems to have explained the perception of material objects by a theory that those objects are at every moment emitting on all sides extremely small images of themselves which strike on the organs of sense. It is to this theory that Aristotle probably alludes when he says (*Metaph.* iv. 5), " Democritus at any rate says that either there is no truth or to us at least it is not evident. And in general it is because these thinkers suppose knowledge to be sensation, and this to be a physical alteration, that they say what appears to our senses must be true."

Such then is the materialistic naturalism of the ancient philosophy. Bodies and souls, objects and worlds, are composed of atoms, the phenomena of nature and the acts of thought are movements of atoms. There is not, never has been, never can be, anything but atoms, void and time, these are the conditions of movement, and movement is the reality of the phenomenal world.

There remained, however, one problem unsolved ; it led to an important and somewhat inconsistent modification by Epicurus of the theory which he adopted. This was the problem of direction. Bodies fall. Their natural direction is downwards. If bodies seem to rise it is either because their fall is relatively slower than that of heavier bodies or, if the direction of their upward movement is absolute, it is due to a rebound from the clash of colliding bodies. Apparently this difficulty was met by supposing that the void is infinite, that atoms are indestructible, that worlds are for ever being formed and unformed, and that their number is infinite. In such a world-view absolute direction could be accepted as fact without introducing direct contradiction. But a new difficulty occurred to Epicurus. If atoms are falling perpendicularly by an inherent natural movement in the infinite void, they will pursue parallel courses from which there is nothing to turn them aside and no heaps or conglomerations will be formed. He introduces therefore a new notion. From time to time he supposes the atoms show a slight inclination, imperceptible and capricious, which Lucretius named their *clinamen*. It draws them from time to time out of the perpendicular and brings them into collision, causing them to form masses. The interest of such a theory, however, is not its physical importance, for in that respect it is quite arbitrary, but that it is inspired by the need of the philosopher to find in nature some basis for the free action of the human soul.

This, then, became the accepted form in which space, time and movement entered into the ancient nature-philosophy. As a philosophical concept atoms and the void could not withstand criticism. On what principle could a limit to divisibility be fixed ? To appeal to perception is impossible for by the hypothesis the perceptible is divisible. Is the appeal to conception any more successful ? Shape or form itself involves the notion of whole and part. It is not difficult indeed to show that the concept of the atom is riddled with contradiction, and moreover possesses no principle by which a synthesis of contradictions can be effected on the Hegelian method. It is a whole without parts. It is a quantity with no quality but its quantity. The void is even more difficult to conceive. It is a pure negation posited as the very basis of reality. It is an absence which forms the absolute condition of presence. Finally, its occupancy supposes a matter which is ultimately indivisible filling a portion of a space which is divisible to infinity.

On the other hand, the atomic theory is not a baseless speculation ; it is grounded in the reality of experience. Moreover, it is not a rough and ready practical solution of an insoluble theoretical problem. It is based on a sound intellectual principle which we may even describe as an intellectual instinct ; the principle that from nothing there is nothing, and the application of this principle to points and instants. Extension is not composed of extensionless points, duration is not composed of durationless instants. The very same intuition which makes the philosopher of mind affirm the moment of experience to be a specious present makes the natural philosopher affirm the atom to be the spatial unit.

The ancient atomic theory has little but a merely outward resemblance to the modern atomic theory.   The latter is not a philosophical theory though of immense philosophical interest. It is a purely scientific theory and not an effort of the mind to conceive the ultimate constitution of matter based on deductions from the logical principle of non-contradiction.   It is scientific in the meaning that it is based on discovery, and that the mathematical formulae by which it is expressed are submitted to the test of experiment, and corrected continually by the results of experiment. It is only in the sense that the atom of modern science is a conceptual object which can never be brought to a direct perceptual test, for the reason that its size is below the amplitude of the waves of light and therefore can never be made visible to our ordinary illumination, that it is permissible to indicate it by the same name.   In contrast to the concept of the ancient philosophy the atom of modern physical theory is not simple and undifferentiated but infinitely complex.   The discovery of the x rays, and their application to the analysis of crystal structure, with the consequent increase in the range of our direct perceptual penetration of matter, have indeed revealed in a positive manner the nature of molecular and atomic structure, and have to that extent given a surer basis to physical science. Yet even this vast extension of perception which our modern world possesses, as compared with the ancient, does not relieve it from the necessity of conception for its idea of the ultimate nature of physical reality.

The value of the ancient theory is that it shows us the form in which space, time and movement provided the scheme of a philosophy of nature. Space was the void, a concept of pure infinite emptiness. Time was the other formal expanse which reality required, but it seems to have been taken for granted and not conceptually analysed as space was. Movement seems ultimately to have been explained by the principle that something is more than nothing and therefore that the something occupying space must, by the very fact that it is something, fall through the void which is nothing. It is clear that to the ancient mind there is one fundamental empirical fact which is accepted as ultimate, apparently unconsciously, and this is the fact which we now call gravitation, and which to them meant the weight which made everything sink in the void.

CHAPTER IV

THE VORTEX THEORY

THE atomic theory of Democritus supported, and indeed for the most part represented, rationalistic and materialistic opinion throughout the whole of the pre-Copernican period. The theory was essentially atheistic, but the reason of its atheism is not immediately evident. It is difficult at first to see why the constitution of nature should have any relation whatever to the question of the origin of nature. As a matter of fact also those who accepted the atomic theory, even Epicurus himself, did not on that account deny the existence of gods. The atheistic character of the theory lay entirely in the fact that its argument dispensed with the *necessity* of God. The ultimate constituents of reality, the atoms, were by their very concept absolute. Creation and annihilation could only have meaning in respect of the grouping of the atoms, the creation or annihilation of an atom would involve self-contradiction. The annihilation of the atom contradicts the notion of a limit to its divisibility ; the creation of the atom contradicts the notion of its simplicity. Clearly then if there are atoms uncreated and imperishable whose combinations (like the letters of the alphabet) produce infinite variety, we have in them a self-sufficient ground of nature. The world may have arisen by chance ; there is no necessity to postulate a creator. So when Dante sees Democritus among the ancient sages in the first circle of the *Inferno*, he refers to him as " Democrito, che il mondo a caso pone," Democritus who ascribes the world to chance.

A curious glimpse is afforded to us of the medieval mind, and the form which materialism and rationalism assumed for it in the scholastic period, in another passage of Dante (Canto X. of the *Inferno*), where he describes the punishment of the heretics. Who are the heretics ? They are not the adversaries against whom Athanasius and Augustine struggled in the formation and interpretation of the creeds, and Dante is some centuries before the Reformation and the institution by the Holy Inquisition of the Auto-da-fe.

The heretics we find are the followers of a materialistic philosophy, not teachers of false doctrine. Dante names them the Epicureans. They include some famous Florentines of Dante's time, and also the Emperor Frederic II., who gathered to his court at Naples and Sicily erudite Grecians and Saracens and revived the classical learning. They lie in their tombs on the fiery plain surrounding the fortified walls of the city of Hell. As Dante, guided by Virgil, passes along, they push up the covering stone of the sepulchre anxious for news of the living. They are the rationalists who thought this life is all, and that the tomb is the end. "There the wicked cease from troubling and the weary are at rest." Alas! they discover that "their worm dieth not and their fire is not quenched."

Apart, however, altogether from the religious and ethical questions involved, the concept of atoms and the void, furnished to pre-Copernican thought the type of physical reality. The void was Euclidean space in its purest uncomplicated form. It was absolute in the sense of perfect emptiness. The puzzling fact in regard to the atoms was what we now call gravitation. It could be determined empirically and its law stated, and for the conglomeration of atoms which forms our earth its direction could be fixed ; but the falling of the atoms in the void was evidently embarrassing. Putting ourselves at the standpoint of the ancient world-view we can see that no means exist to decide whether the atoms are falling continuously through eternal time in infinite void or whether they are at rest. If, however, it is not in the nature of the atom to fall in the void it is difficult to understand why there is movement anywhen or anywhere. Movement would have to be impressed from without and the ground of the self-sufficiency of the atomic theory would be gone.

The whole of this ancient world-view was changed by the Copernican discovery. This discovery brought about a most profound and complete revolution in human thought, turned science and philosophy in a new direction, and with a new world-view opened to the human mind new problems, new methods and reformed concepts. No such tremendous effect in determining the intellectual development of our race has approached in importance that which followed this discovery. If we would classify scientifically the historical stages in the evolution of philosophy and distinguish them by a central epoch we ought to mark all theories as being pre-Copernican or post-Copernican. That is in fact what we do when we name Descartes as the founder of modern philosophy, for Descartes was the systematizer of that discovery, the philosopher of that revolution, and his principles, his method and his system are completely determined by it.

Yet the Copernican discovery seems a simple enough matter and we are generally inclined to wonder how it could have been possible for mankind to continue so long without someone suspecting that the celestial movements were an appearance consequent on our own translation. We understand the shock to the religious faith of those who had pictured this earth as the scene of a tragedy, prepared from eternity by the divine source and sustainer of the universe, and for whom human history led up to, and followed from, that unique event. But so easily have we come to adapt ourselves to the new world-view that we are unconscious of the change, and indeed our difficulty in reading the ancient philosophers is to remember that their concepts were concerned with an imagery totally different from ours.

Suppose one had been born in a smoothly running railway-carriage, and brought up to find in it all the conveniences and necessities of life, able to look out on the world through which it is continually journeying. It would and it must seem to one so circumstanced, that the panorama without is in ceaseless movement. Every alteration of relative position of the moving system would appear as a movement of the panorama. This was the condition of the human race. It developed intellectually through continuously successive ages without ever suspecting that the movement it looked out upon in the panorama of the heavens might be an appearance due to its own translation. The discovery came suddenly and with something of a shock. But the discovery having been made, the evidence for it accumulated with such force that the world-view adapted itself to it, and we are no more able to-day to return to the old world-view than we are able when we take a railway journey to believe that our carriage is at rest and the landscape moving.

The philosopher of this new world-view was Descartes. It is no mere chance coincidence that Descartes was philosophizing and elaborating a

new system of the universe when Galileo was
experimenting to prove the earth's movement.
A new concept of truth and reality on which the
new aspect of the universe could be rationalized
and harmonized was imperatively called for. It
must be a return to absolutely first principles.

Descartes laid down two principles of philosophy, one subjective and one objective, and both
the direct outcome of the Copernican discovery
and its revolution in the world-view.

The first principle is that the intellect alone by
the clearness and distinctness of its ideas can
furnish a criterion of truth. The senses are
deceptive, the source of confused and obscure ideas.
The senses do indeed induce belief, and seem
to furnish an assurance of truth, but their purpose
is not to lead to truth but to preserve the body.
It is then not to sense but to the clear and
distinct ideas of the mind, to reason, that we must
turn for true knowledge. Why the clear and
distinct ideas of understanding should possess
superiority over, and greater validity than, the
obscure and confused ideas of sense was indis-
coverable in their nature. Descartes fell back
on the proof of the existence of God and the
impossibility of our conceiving that in the case

of clear and distinct ideas God could deceive —a principle which no longer appeals to us. On this distinction between sense and intellect was founded the well-known method of Descartes. He proposed to doubt everything which could possibly be doubted in order to discover, as a starting point, some fact which expressed, in the clearness and distinctness of its idea, a truth which it was not possible to doubt. Such fundamental truth he claimed to have discovered in the famous *Cogito ergo sum*. It is easy to see the connexion of this with the Copernican discovery. Had not that discovery clearly demonstrated that mankind, universally and continually, trusting the interpretation of direct sense experience, had lived in age-long error ?

The second principle of Descartes concerns the objective reality of the universe. The universe is a mechanistic not a materialistic system. It is not the outcome of the behaviour of atoms in a void, it is the mechanical disposition of matter resulting from the imparting to it of movement. The essence of material substance is extension alone and there is no void. Movement is not change of place but relative change of neighbourhood. Movement is only possible in a plenum ;

movement in the void is unmeaning and self-contradictory. Movement in a plenum is a vortex movement, that is, a movement which involves simultaneously every part of the system and is not propagated from part to part. The universe is a system of vortices, each vortex determining vortices within it, and determined by relations to vortices without it. The solar system is a vortex, the fixed stars are similar vortices, and the planets and their satellites are all vortices within the vortex, and all movement down to the beating of the heart and the circulation of the blood is one in principle, having its part in the universal mechanism. This constitutes the first great systematization of the universe in accordance with the revolution in astronomy.

These two principles, the subjective principle or new method and the objective principle or mechanistic interpretation, have had a diverse fate in the history of thought. While the first has been accepted as marking the beginning of a new period of philosophical speculation so that we regard Descartes as the founder of modern philosophy, the second, the cosmological and physical theory, is neglected and forgotten, or read, when it is read, as an intellectual curiosity

E

with no relation to present physical or metaphysical science. But in Descartes's own time and during the development of Cartesianism in the half century which followed, there was no such dissociation of metaphysics and physics, philosophy and science. It was the vortex theory which established the fame of Descartes.

" Give me matter and movement and I will make a world," was the famous challenge which he threw down to the theory of atoms and the void. Keeping in mind that for Descartes matter is extension, we can translate it to mean that the variety and the uniformity of the universe are a function of systems of movement. To understand it we must examine a little more closely the three distinct doctrines, interconnected and interdependent, on which it rests. These are (1) that the essence of matter is extension ; (2) that movement is relative not absolute, not change of place in an independent expanse but the relative change of neighbourhood of extended systems ; and (3) that nature is a plenum, there is no void and movement in a plenum is a vortex.

The first of these doctrines is fundamental. It is the ground of Descartes's rejection of the void. Extension is not the empty place in which there is or is not matter, it is the essential attribute of matter. The direct argument for this is that extension is the only attribute which is inseparable and indistinguishable from material substance. Every other attribute—colour, weight, sonority, resistance, shape—can be thought absent, but if we abstract from its extension, material substance itself is annihilated. The apparent contradiction that the extension of any matter is variable, as instanced by rarefaction and condensation, is easily explained as the disposition of a matter's extension in relation to other material extensions. Extension being the essence of material substance, if and when matter moves its extension moves. Extension is not the quantity of emptiness matter fills. To say of empty space that it is extended is to endow it with the essential attribute of material substance, and so to deny that it is a void. The rejection of the atoms is still more direct. They are geometrically impossible, not on account of self-contradiction in the concept of indivisible particles having form or shape but no parts, but on account of their unchangeability. Movement would be impossible if the constituents of matter had unalterable shapes.

The second doctrine concerned the relativity of movement. The Copernican theory had merely substituted a heliocentric for a geocentric astronomical hypothesis. Descartes saw that it raised the metaphysical problem of movement. Nothing is at rest in the whole system of nature if being at rest means being in a moving system and not being carried along in its movement. But this is not what we mean by rest in ordinary experience. We say we are at rest and not moving when the members of our system keep their relative positions notwithstanding that the whole system may be in movement of translation or may be itself not moving but borne along in a movement. We are at rest, for example, in the cabin of our ship when wind and stream are transporting the ship to France. The earth may be considered at rest if we mean that it is carried along its path through the solar system like a ship on the ocean. It is no longer possible then, Descartes argued, to regard anything as in its nature at rest. There are no fixed immovable points. Nothing has a permanent place except in so far as it is fixed by our thought. The common notion is that a body moves when it changes its place in a void. Strictly defined, movement is not change of place but change of neighbourhood, it is the translation of a part of matter or of a body from the neighbourhood of bodies with which it is in contact into the neighbourhood of others. We can only define it relatively. When I push a boat off a beach it is merely convenience which makes me express it as a movement of the boat relatively to the beach at rest and not as a movement of the beach relatively to the boat at rest.

The third doctrine had for its main argument the defence, against the atomists, who denied it, of the concept of the possibility of movement in a plenum. The argument of the atomists had been that there must of necessity be the void, for without it movement is impossible. Where, they asked, is the place into which to move if every place is already occupied ? Movement, Descartes contended, is possible in a plenum if the chain of moving members is complete so that the last of the series moves into the place of the first. Such movement is really changing place and not passing through a void which does not change.

Indeed, if in moving we did not carry our extension with us, how could we have a science of geometry? In geometry we are not measuring vacuum, we are measuring extension. The

figures we construct and study in geometry—
circles, squares, triangles, cubes, spheres—are
measurements and constructions of the extension
which moves with the earth, not of a supposed
vacuum independent of that movement and
indifferent to it.   Endless complexities and contradictions, actual as well as logical, arise if we
attempt to interpret geometry in terms of vacuum.
We say, for example, that moving is the opposite
of resting.   Now suppose that, following the
common notion, we define movement as change of
place, and rest as remaining in the same place,
then we see at once that for anything on the earth
to be at rest, it must be parting company with the
earth at a prodigious velocity.   Descartes had
therefore the choice of two alternatives.   Either
extension is an attribute of material substance and
accompanies it in all its changes, or it is vacuum
existing independently of substance.   If he chose
the latter he must sacrifice geometry, for no means
exist of measuring vacuum.   His philosophical
theory, though opposed to the universally accepted
notion, was a necessity of thought and a great
advance in mathematical and physical method.
When I move about a ship I am really moving
notwithstanding that to the observer on shore I

may be at rest ; the beating of my heart, the revolving wheels of the watch in my pocket, are regular movements, though the tracing of them on a chart, against an absolute background, would be hopelessly complicated and different for different observers.

The application of this principle led to the construction of the magnificent scheme of celestial or rather cosmical mechanism which amazed and held spell-bound the intellectual world of the latter half of the seventeenth century—the vortex theory. The solar system is a vortex with the sun as its centre, extending beyond the orbits of the distant planets. It is bounded by other vortices. These are the fixed stars which like our own sun are centres of revolving systems.

There are two laws of nature which Descartes formulates. They are rational deductions empirically verifiable. The first is that everything remains as it is till something changes it. The second is that every body which is moved tends to continue moving in a straight line. The rationale of the second law is that the straight line being the shortest line measures the force. The first law explains how bodies get involved in vortex movements. By these two laws he accounted for the planets in the solar vortex.   The planets, he said, are at rest so far as their sky is concerned. They are not careering through space but being carried along with the moving system, that is, the solar vortex.   They have been caught up in it, projected it may have been from other vortices and sent travelling in a straight line until they became involved in our system.

Such is the mechanism which Descartes substituted for the old materialism.   I have not dealt with details but tried to bring into relief its essential features.   It was the constructive work of a single genius.   It enjoyed a brilliant vogue, capturing the imagination of more than one intellectual generation.   Yet it has passed away so completely that it is hardly remembered even as a stage in the `evolution of scientific theory. The picture of the physical universe as a system of vortices, described with such mastery of minute detail, and with such assurance, in the *Principia*, is no doubt as far removed from our present imagery of physical reality as the description of the organism controlled by the animal spirits in Descartes's *Les passions de l'âme* is far removed from our modern physiological concepts.   Nevertheless, in the one case as in the other there are important principles insisted on from the neglect of which science has distinctly suffered. In outward resemblance Descartes's world-view is extraordinarily like that which is presented to us by the general principle of relativity. So much so that it appears at first as though in rejecting Newton's concepts we are simply returning to those of Descartes. We have only to remember however that the whole development of physical science has, in recent times, come to centre round the electro-magnetic theory, and that this concerns a continuity of experimental discovery in a realm of phenomena entirely unexplored by the mathematicians, astronomers, and mechanicians of the seventeenth and eighteenth centuries, to see that there can be no simple reversion. The principle of relativity is in reality the rationalization of the electrical theory of matter. It is interesting to note how it was anticipated in the principles which suggested to Descartes the vortex theory. The concept of the vortex is itself a quite striking anticipation of the modern concept of the " field of force."

Descartes distinguished two ideas as intellectual and therefore not subject to the deceptive appearance which characterizes the ideas dependent on sense perception. These are thought and extension. By these concepts he distinguished the essence of the two substances which we refer to familiarly as mind and matter. So far as physical theory is concerned the important concept is extension. In Descartes's theory that extension is the essential attribute of matter it is denied that there is any void or pure space within which matter moves. Extension is not something moving matter leaves behind it or of which it exchanges one quantity for another. The moving mass or system carries its extension within its movement. From this it follows that all movement is relative and concerned only with the relations of material systems to one another. This accords completely with the modern theory of relativity.

It is curious that the duration of the universe did not impress Descartes as having anything like the importance which he attributes to its extension. He recognizes that the universe endures, but it is not, he thinks, by reason of anything in its essence, the fact of duration simply shows its dependence on God. If duration is the essential attribute of a substance, we must conclude that this substance is God, but Descartes does not himself draw this conclusion. Time is as necessary as space for a mechanism. Without time the machine cannot work, but time plays the part of independent variable, it has no grip on the reality of the machine, any more than the time the clock measures is part of the contrivance. To us extension and duration are correlative and inter-dependent. From the historical standpoint this is of peculiar significance. The great Copernican revolution brought in a new concept of the celestial mechanism, and incidentally it re-formed our view of the spatial universe. It was not until three centuries later that a reform of our concept of the duration of the universe, parallel to the Copernican concept of its extension took place. It followed the great biological discoveries of the nineteenth century. The Darwinian theory brought as complete a revolution in our conception of time as the Copernican theory had produced in our conception of space. To Descartes, therefore, duration appears not as the essence of the universe, but only as that which is necessary to its existence. Its continuity from moment to moment depended on a creating and sustaining power.

It is, then, in the concept of matter as extension, and in the concept that movement and rest are mutually dependent on systems of reference, and in the concept of translation as relative to the members and parts of the system, together with the rejection of any absolute zero, such as the void, affording a standard for the measurement of absolute velocity, that Descartes's vortex-system anticipates the principle of relativity. The doctrine that material substance consists in extension alone does not mean that pure extension exists materially without any other quality or character whatever. It means that extension is constant and that no other attribute of matter is. Any other attribute which matter may have, or any attribute that it may need in order that we may apprehend it, is variable. It follows that all the diversity and endless variety of the material universe must be due to movement and a direct function of movement. This follows simply from the fact that extension is not stuff and therefore cannot harbour occult properties—essences or forces.

CHAPTER V

THE PROBLEM OF GRAVITATION

It is a curious thing that Descartes who proposed a new method of philosophy, the distinctive form of which is universal doubt, and the principle of which is that nothing must be accepted as true unless its evidence is presented to the mind with a clearness and distinctness which excludes doubt, should have worked out a hypothesis of the system of the universe complete down to the minutest detail. There is no greater contrast in the history of western intellectual development than his system presents to the method and philosophical system of Newton which completely supplanted it. Newton's method was experiment, and his philosophy he described as the experimental philosophy. This does not mean that Newton himself was an experimentalist. He conducted no investigations in the manner, for example, of Galileo. He was a mathematician like Descartes, and all that he insisted on in his natural philosophy was that nothing should be accepted as true on the clearness and evidence of the pure idea, unless it had first been submitted to the test of positive experimental proof. It is also curious, in comparing these two great minds and the work they accomplished, to observe that while the speculative philosophy of Descartes has secured a permanent place in literature, his physical system, which was the supremely important thing to his contemporaries and his successors, is entirely rejected and studied only, if it is studied, for its antiquarian interest. Newton, on the other hand, who was equally famous to his contemporaries and immediate successors as a speculative philosopher, is now remembered as a great mathematician and as the discoverer of the universal law of gravitation, while his philosophy is entirely neglected. Even our knowledge of his great discovery is not first-hand. The *Philosophiae Naturalis Principia Mathematica* is not a classic which takes a place with Hobbes's *Leviathan*, Locke's *Essay on the Human Understanding*, Spinoza's *Ethics*, and the other great works of his contemporaries, it is a concealed book for all but advanced mathematicians, and our knowledge of Newton's discovery is enshrined in a moral tale concerning his reflection on the fall of the apple. It is strange there should be so little direct study of the work of one of the greatest geniuses our country has produced. It is further of interest to note that though Newton in his life-time won immediate and unquestioned recognition—he was elected President of the Royal Society twenty-five years in succession—his views never obtained wide or important acceptance. It was after his death (1727) and in consequence of the publication of Voltaire's *Elemens de la Philosophie de Newton* in 1738 and the translation of the *Principia Philosophiae* into French some years later that Newton's " Philosophy " triumphantly deposed the Cartesian vortex theory and became the accepted basis of physical science.

Newton was born in 1642, and was therefore eight years old when Descartes died, and his years of study and research were those during which the Cartesian system reigned unchallenged. The *Principia Philosophiae*, the work of many years, written in Latin and bearing the same title as Descartes's famous work, was published in 1686-7. It presented his theory of gravitation and his formula of the universal law. The famous story that this was a sudden discovery following a meditation set going by seeing an apple fall in the orchard of Woolsthorpe (from a tree which was shown to visitors for about a century, and which after it had fallen through decay was carved up into souvenirs), is generally rejected as legendary and mythical.   Certainly if we suppose, as is usually implied, that Newton at the time was of a poetical and impressionable turn of mind and awakened by a simple occurrence to the thought of the mystery underlying natural processes, nothing can well be more improbable than the story. When, however, we take the anecdote in its historical setting we see at once that, whether it be legendary or not, something very like it must have happened.   What is certain is that the circumstance, falling apple or whatever it may have been, did not originate the meditation but broke in upon it.   The authority for the story is Voltaire, who had heard it from Madame Conduit, a niece of Newton, married to a Fellow of the Royal Society who was one of Voltaire's intimate friends.   The story is :   " One day in the year 1666, Newton in the retirement of the country " (he had withdrawn from Cambridge to Woolsthorpe on account of the plague), " seeing the fruit falling from a tree in the orchard, according to the story which his niece Madame Conduit told me, fell into deep meditation as to the cause which thus draws all bodies in a line which if prolonged would pass almost through the earth's centre. What, he asked himself, is this force? It acts on all bodies in proportion to their mass and not to their surfaces ; it would act on this fruit now falling from this tree were it so removed that it had three thousand or ten thousand fathoms to fall. If this be so, then this force must be acting from where the moon is right to the centre of the earth. If so, then, whatever this power be, is it not the same as that which keeps the planets moving round the sun and the satellites of Jupiter in their orbit round that planet? Now it has been shown, by the inferences drawn from Kepler's laws, that these secondary planets are weighted towards the centre of their orbits, more in proportion as they are near, less in proportion as they are distant, that is reciprocally according to the square of their distances. A body in the moon's position and a body close to the earth must both weigh on the earth in exact accordance with that law."

F

This story shows clearly the subject of Newton's meditation. He was pondering over Descartes's system of celestial movement. But why should that be disturbed by the falling apple ? It flashed across him that here was a phenomenon in flagrant violation of Descartes's vortex principle. By every reason alleged by Descartes the apple ought to have flown upward and outward and not to have fallen downward. The centrifugal force of a vortex causes the heavy body held by it, when released, to leave it at right angles to the axis of rotation, as when for example the stone in a sling is released. This at once brings to mind the laws of Kepler, in particular the third law, that the cube of the distance of each planet from the sun is proportional to the square of its time of revolution. But what strikes Newton is that weight must be determined by mass and distance, and not, as Descartes's principle required, by surfaces in changing relations consequent on movement. It is quite certain therefore that if it was not a falling apple it was some analogous circumstance which originated in Newton's mind the doubt concerning Descartes's principle and led to the formulation of the new law and to the return to the concept of absolute space and time as the basis and framework of the physical universe.

To make this clearer it will be well to look a little closely at the way in which Descartes conceived the vortex to have worked in bringing about the distribution of matter in the solar system. In rejecting the concept of atoms and the void Descartes had no need to endow matter with weight as an essential attribute. As there is no void, and therefore no distinction between space and matter, that is to say, as matter is not conceived as in space or space as containing matter, the one and only essence of matter is extension. Without movement this is pure homogeneity. If we conceive matter at rest we must of necessity conceive its parts, if it consist of separate parts uniform in character, as regular figures, such as cubes or hexagons, for only such will fit together to compose a plenum. Suppose now that movement is originated, it must, on the theory, involve relative change throughout the mass and the effect must be that the shape of the figures will alter and continue to alter till they become entirely spherical. The more and the finer the spheres become the more easily will they pass one another in moving and their velocity will tend to increase continually. There will eventually be formed very minute and perfectly spherical particles, a very fine dust through which the particles will cleave an easy way, and denser particles which will be driven outwards and form a dust on the circumference of the vortex sphere. This, according to Descartes's natural history of the world, is what has actually happened in our system. The fine spherical matter has collected at the centre, it is the luminous or fire element and composes the sun, the finer dust fills the firmament and is the air element or transparent matter, lastly there is opaque matter which composes the crust of the earth. The place of this matter is somewhat unsatisfactorily explained in the system of Descartes. He suggests that the sun spots are a kind of scum which collects on the surface of the mass of spherical matter, forming the denser masses which are thrown off and become celestial bodies travelling in a straight line until they are captured by some vortex into which they enter. He supposes that our planets may have come from other systems and been caught up in the solar vortex. And he also supposes that the earth is itself a vortex within the solar vortex with a similar structure, the fire element at its centre, the air element surrounding this, and the opaque crust on the circumference. The obvious difficulty of this theory, which Descartes never saw, and which is only obvious to us since Newton's discovery, is that according to it the heavier the matter on the circumference the greater should be its tendency to obey centrifugal force and fly off. Descartes's theory therefore offered an explanation why the planets do not fall into the sun, but it had no explanation to offer as to why they keep a constant position in the solar vortex. It could explain why movement is centrifugal or centripetal at the equator, not why it is radial to the centre from every point on the circumference. And lastly, it made no attempt to explain the behaviour of heavy objects on the surface of the globe.

Let us now turn to Newton's theory of gravitation. There is an active power which impresses on all bodies a tendency to move towards one another. In the celestial bodies this power acts in inverse ratio of the squares of the distances of the centres of the masses and in direct ratio of the masses. The power is called *attraction* when we refer to the centre, *gravitation* when we refer to the bodies which fall to the centre. It is this power which causes bodies, liberated on the surface of the earth, to fall to the earth in direct line to the centre. The same force probably acts on the light emitted by luminous bodies, but we are ignorant in what proportion and have no means of discovering. It was not, however, till long after Newton had conceived the idea, that he was able to confirm the theory and formulate the universal law. The account of his difficulties, which Voltaire has given us, is of special interest for the light it throws on his character as well as on his method. To determine the amount of the attraction of the earth on the moon he must know the radius of the earth and the distance of the moon. Here is Voltaire's account. " Newton was at the disadvantage that the only data he possessed on which to base his measurement of the earth were the faulty calculations of English pilots who reckoned sixty English miles to a degree of latitude, the true amount being more nearly seventy. And yet there had been a more correct measurement of the earth. An English mathematician, Norwood, had in 1636 measured fairly accurately a degree of the meridian and had discovered it to be about seventy miles. But

although this measurement had been made thirty
years before, it was unknown to Newton.   The
civil wars which had afflicted England (always as
disastrous for science as for general prosperity),
had buried in oblivion the only exact measurement
of the earth which then existed, and there was
nothing available but the vague estimate of the
pilots. Employing this measurement it was
found that the moon was too near the earth and
the expected equations did not come out right.
Newton did not try to supplement his theory by
forcing nature to accord with his ideas.   He gave
up his great discovery, notwithstanding that the
analogy with the other stars had appeared to make
it so probable, and though it seemed to come so
little short of demonstration.   This example of
good faith alone deserves to give great weight to
his opinions."

Later on more exact measurements were made
in France giving twenty-five leagues as a degree
of the earth. This gave the distance of the moon
as sixty radii of the earth and was exactly what was
required to give Newton the demonstration of his
theory.

Newton's theory is that the moon is travelling by its own inertia in a straight line and at the

same time being attracted by the force of gravity towards the earth's centre, the influence being mutual, the composition of the forces giving the moon's orbit. By applying Kepler's formula he could calculate the exact amount of the gravitational force, or the weight of the moon on the earth at its distance from the earth's centre.

Why was this theory unacceptable to the Cartesians ? The main reason is clear. It proposed the very thing which to their principle was most abhorrent. Gravitation was an occult influence, and the Cartesians would accept no explanation which was not wholly intelligible and explicable by the aid of simple mechanics. Further, it raised the whole problem of action at a distance. According to the Cartesians a body might exercise pressure on another body, and in fact this was a constant phenomenon, but in every case the propagation was instantaneous by reason of the continuity of matter in the vortex. The bell rings at the same moment at which the cord is pulled. In this way the Cartesians explained how the sun, though distant, could be the source of the sensation of light. No emanation passes from the sun to our organs of sight, but a pressure is exerted and propagated through the transparent matter of the firmament. In like manner they explained the tides. The pressure of the moon, propagated through the intervening medium, depresses the liquid element on the surface of the earth causing it to bulge in the tidal wave. Action at a distance was puzzling and even disconcerting to Newton, and although if it take place in fact it has to be accepted, on the principle that nothing is impossible to God, yet he inclined to the hypothesis that the space, in which the masses in the solar system move, is pervaded with a subtile substance, an ether, through which the influence of attraction and gravitation is conveyed.

It will be seen then that Newton's discovery demanded a complete revision not only of the special hypothesis of the Cartesian vortex but of the whole philosophical concept on which the Cartesian mechanism was based. Attracting and gravitating masses, forces determined by the inverse ratio of the squares of the distances between the centres of the masses, were clearly incompatible with the concept of a material substance whose essence was extension alone and whose form was determined by relative movement within a self-contained and self-sufficing system. The new system required a framework of absolute space and time—space which was independent of, and indifferent to, the masses acting on one another within it, and a time flowing indifferently to the changes which it measured.

Newton therefore takes us back to the old atomic theory. He re-affirms the void and a matter which is atomic in the sense that it may be impressed with movement (for to deny the atom would be in effect, as we have seen, to deny the possibility of movement). But it is not a simple return. There is a notable advance in the fact that the theory of gravitation has completely overcome the ancient difficulty regarding absolute direction. The problem which exercised Democritus and produced Epicurus's arbitrary hypothesis of an " inclination " is replaced with this law of universal gravitation. The mysterious force of attraction, which observation and experiment have required us to accept, remains a mystery in natural philosophy.

From the standpoint of physical science the gain seems enormous. The whole scheme of the universe is simplified and settled on a basis of common-sense. We are provided with absolute standards of reference and appear to have no obstruction to the unlimited advance, through experimentation, of exact knowledge. But the new concepts raised very serious difficulties in philosophy. To Newton himself the materialism of his system presented no difficulty. He held firmly to the necessary existence of God, and the creation or annihilation of matter seemed to him to come easily within his conception of the power of God. The old difficulty of the atomists that matter in its very concept was a necessary existence, and therefore eternal and indestructible, did not trouble him. " It seems probable to me," he says, " that God in the beginning formed matter in solid, massy, hard, impenetrable, moveable particles ; of such sizes and figures, and with such properties, and in such proportions to space, as most conduced to the end for which he formed them." But space and time themselves—what was their relation to God ? All existence depends upon them and they depend on nothing and their non-existence is unthinkable. At the close of his questions in the *Optics* he says: " Do not these phenomena of nature make it clear, that there is a being incorporeal, living, omnipresent, who in infinite space as in his *sensorium* sees, discerns and understands everything most intimately and with absolute perfection ? "

By the term *sensorium* Newton expressed his idea of the relation of God to infinite space. God is not in space and space does not contain God, it is part of God's nature. It was a difficult, and ambiguous expression and of course not allowed to go unchallenged. It was the subject, of a correspondence between Leibniz and Newton's most famous disciple, Clarke, in 1715-16, a correspondence which was only interrupted by Leibniz's death. What Newton intended by the doctrine was that allowing, as we must, for the impossibility of expressing God's nature in human terms, we have to admit that nothing can act, know or see where it is not, and when therefore we affirm of God that he is omnipresent we mean that he acts, knows and sees at every point and in all points of space. Space therefore is God's sensorium. Leibniz criticized this as being tantamount to making God's relation to the world analogous to the relation of the soul to the body. The sensorium would then represent in God's nature what the pineal gland represented in human nature, in Descartes's theory of the relation of the soul to the body. The real difficulty for Newton was, that however he might wish to conceive God's relation to nature, the absolute existence of space and time as physical realities was fundamental for his conception of gravitation as the attraction of masses. Space is not where God places masses, and time is not when God elects to create them, for then we should have to say that if God had chosen a different place or a different time for creation, the different places and times would have been not different but identical, and this involves absurdity. Physical science, however, has ignored these philosophical difficulties and framed its concepts on the supposition of absolute space and time, and this pure assumption has come to be a common-place of every day thought.

To go back to the distinctive work of Newton and the formulation of the universal law of gravitation as a consequence of his discovery, great light is shed on his own conception of the nature of his work and on its relation to philosophical theory by the closing passage of the *Principia Philosophiae* which I will quote. It also brings out the beauty of his personal character.

"So far I have expounded the force of gravitation by celestial phenomena and by those of the sea, but I have in no way attempted to assign the cause.

That force comes from a power which penetrates to the centre of the sun and of the planets, without any diminution of activity ; and it acts not in proportion to the quantity of the surfaces of the particles of matter (as mechanical causes do), but according to the quantity of *solid* matter ; and its action extends on all sides to immense distances, diminishing always in exact ratio according to the square of the distances."

" I have not tried to deduce the cause of these properties of gravitation from the phenomena and I make no hypotheses. For whatever is not deduced from phenomena is a hypothesis ; and hypotheses, whether they are metaphysical or whether they are physical, whether they presuppose occult qualities or mechanical qualities, have no place in. *experimental philosophy.* In this philosophy propositions are deduced from phenomena and general propositions are obtained by inference. Thus the inpenetrability, the mobility and the impetus of bodies and the laws of their movements of gravity have been set down. And it is shown that gravity really exists, and that it acts according to the laws I have expounded and that it applies to all movements of the celestial bodies and of our sea."

" It is not now possible to add anything concerning the very subtile spirit pervading heavy bodies and latent in them ; by whose force and actions the particles of bodies are mutually drawn together at minimal distances, and the contiguous cohere ; and concerning electrical bodies acting at greater distances, now attracting now repelling neighbouring bodies ; and how light is emitted, reflected, refracted, inflected, and also how it warms bodies; and how all sensation is excited and how in animals the limbs are moved by volition, to wit, by the vibrations of this spirit propagated through the solid threads of the nerves from the external organs of sense to the brain and from the brain to the muscles.   These cannot be expounded in a few words, and at present there are not sufficient experiments by which to determine accurately and demonstrate the laws of action of this spirit."

CHAPTER VI

LEIBNIZ AND THE THEORY THAT SPACE IS THE ORDER OF COEXISTENCES

THE effect of Newton's discovery of the law of gravitation was to reinstate the old theory of atoms and the void in the scientific conception of nature-philosophy. Newton could offer no explanation of gravitation, but he definitely established the fact. There actually is an influence or force in masses, whatever their nature, which draws them out of their path, whatever its direction, and whatever the velocity of their movement in it, and this force is proportional to the distances of the masses. But though it restored the old concept of atoms and the void it seemed in doing so to supply just that principle, ignorance of which had constituted the gravest defect in the ancient theory. Instead of the purely fanciful theory of the *clinamen* proposed *ad hoc* by Epicurus, we now have a principle of attraction which however mysterious, can be formulated as an invariable law, and tested by actual experiment. It lent itself to theological interpretation also, for in giving matter weight God was endowing the creation with the principle of its order and arrangement.

The success of the new physical discovery in upsetting the mechanical theory of the vortices was complete. The vortex movement is inconsistent with the gravitation phenomenon in two distinct particulars. The vortex will explain gravitation at the equator, for there the centrifugal line of force is at right angles to the axis and passes through the centre, but if the revolution of the sphere round its axis is the cause of this force, then nowhere else but at the equator should the falling body be directed to the centre, and also weight if it be centripetal force should decrease to zero at the poles. The vortex theory therefore is inconsistent with the fact that a body falling freely towards any point on the surface of the sphere pursues a line which if continued would carry it to or near the centre.

A second fact was also plainly inconsistent with the vortex theory, viz. the fact that the planetary orbits are ellipses and not circles. The vortex will explain circular movement but not eccentricity.

A theoretical difficulty was also pointed out in Descartes's theory of the origin of motion and the possibility of its origin and development in a plenum. Matter at rest according to this theory must, if movement is to be possible, be conceived as disintegrated. Being a plenum it will consist of closely packed figures with plane surfaces, let us say cubes. Now Descartes supposed that movement set going in this plenum would cause the cubes to become spherical by gradually wearing off the angles. But how could this process start without first creating a void? The movement of the cubes cannot alter their relative position without creating void. Before the movement alters the shape of the cubes by fracturing the angles, it must cause their displacement, and the slightest displacement destroys the plenum.

The new philosophy, as it was called by contemporaries, rapidly and completely superseded the old. It did not correct it or supplement it. It did not borrow and incorporate some of its principles while adding new ones of its own. The *Principia Philosophiae* of Newton became what the *Principia Philosophiae* of Descartes had

been. It came to pass that Descartes died and
Newton reigned in his stead. The new principle
was an active force resident in masses, but measur-
able in its effects and capable of developing a
complete celestial mechanism. Nature was a
system of forces residing in masses, distributed
in an infinite and absolute void. The forces
manifested themselves by mutual influences
causing the masses to move and successively
change their relative positions in an even
flowing, absolute, time.

While Newton was working out the great
discovery which was to give a physical system
destined to supersede the apparently firmly
established system of Descartes, a contemporary
philosopher and rival mathematician was opposing
the philosophy of Descartes on purely metaphysical grounds. Leibniz (1649-1716) con-
centrated his criticism on Descartes's conception
of substance and in particular strove to present
a rational theory of the relation of the soul to the
body, of mind to matter, of God to the world.
It is especially in their concepts of the relation
of God to the world, that the philosophical
principles of Leibniz and Newton are antagonistic,
and on this point Newton was most sensitive.

Newton did not recognize any theological difficulty in his nature concept such as that which seemed to make the old atomism essentially materialistic and atheistic. He could find no ground for regarding God's power over matter as limited to its disposition, God could create it or annihilate it at his pleasure. But what was God's relation to an infinite and absolute space and time ? Were they not in their very nature limitations ? The only alternative was to make space and time the nature of God, and Newton took refuge in the attribute of omnipresence. Wherever God is acting he is. This was the meaning of the theory that space is God's sensorium.

The theological problem which seems to us to engage almost exclusively the philosophical speculation of the seventeenth century has so changed its form in our philosophy to-day that we incline to regard it as an outgrown mythology. It is, however, just as vital an issue to-day. All that we have discarded is the anthropological concept, but in the issue between idealism and naturalism we have kept all that was essential in the old theistic problem. The challenge of the old theology to physical and metaphysical theories was whether

God was conceived in them as a possibility or as a necessity. Both Newton and Leibniz believed in God, believed not only that a supreme and infinite being exists, but that the necessity of such an existence can be deduced from the fact of the universe. But while Newton argued from what the universe is to what God's attributes must be, Leibniz argued from what God is to what the universe must be. Newton was convinced that there is a living God, but no more than Laplace, who a hundred years later carried out and developed his principles, had he any need of that hypothesis. Leibniz on the contrary could not move one step without it.

Voltaire, who spread the fame of Newton's discovery and who, as I have tried to show, did more than anyone to secure its triumphant acceptance, pointed his advocacy by the wit and mirth-provoking satire with which he treated Leibniz. Leibniz is the Dr. Pangloss of *Candide*. Voltaire knew well the intellectual strength of the philosopher he satirized. When Leibniz argued against the void that the concept of such a reality was inconsistent with the attributes of power and goodness in God, for it would imply a shortcoming or defect in creation ; that it would imply room for a possibility of creation which had been unfulfilled and would mean that the world was not the best possible ; the argument was received with laughter. To-day we find it difficult to think it could ever have been put forward in seriousness. Yet a study of Leibniz's thought will show how his whole concept of the monads rests upon it. It is his principle of continuity. The world God has created is not a patchwork of stuffs but a city of active workers. The creation is not an abiding-place for souls to dwell in, or a stage whereon to display their activities, the world is souls, and only souls exist. In the perfect city every citizen realizes his own individuality and in so doing fulfils the higher life of the community, the compound or composite existence. To suppose a void then is to suppose that something essential to the perfection of the system is left unprovided for or unfulfilled. This is inconsistent with the idea of a perfect God. Voltaire's biting satire of Dr. Pangloss, preserving his philosophical conviction unshaken when overwhelmed with accumulated disasters, made Leibniz appear a kind of Don Quixote in the popular philosophy of the eighteenth century. It is evident enough to us that whatever force the satire may have against particular theological or even religious theories, it has absolutely none against the conception in its scientific and philosophical application.

Leibniz inaugurated the most momentous movement in modern philosophy. He is the founder of modern idealism. If we divest his doctrine of the peculiar theological form in which he clothed it we see the origin of his theory in the profound dissatisfaction with the materialist attempts to give a rational explanation of the universe. The revolution which his concept of substance marks, is to some extent analogous to the revolution which Newton's discovery produced in natural philosophy. It substituted the concept of a dynamic for that of a static reality, it replaced the notion of stuff with the notion of force. The atoms of Leibniz are forces, activities, just as the masses of Newton attract and repel. But there the analogy ends. No greater contrast is presented in philosophy than the divergency of these contemporary philosophers.

In one instance only they appeared as rivals, and on one side at least the recriminations were bitter. This was the dispute concerning the differential calculus which each claimed to have discovered. When Leibniz published his work he was charged with having stolen it from unpublished writings of Newton which he was known to have seen, and Newton gave substance to the charge, which he evidently believed, in an innuendo in the Preface to the *Optics*. Leibniz resented the charge bitterly and defended himself against it. The sad thing is that one generous word from Newton would have closed the matter, for Leibniz addressed to him a personal appeal. The letter remained unanswered.

Leibniz's philosophy was a doctrine of the true atoms of nature, the mode of their activity, and the way in which they combine to form the universe. The old atomic theory stood condemned by its unsolved contradictions and constituted therefore a continual offence against human reason. The fundamental principles of the intellect are the law of identity and the law of sufficient reason, and both are flagrantly violated in the theories of atoms and the void. Divisibility is part of the concept of extension. In declaring anything extended we are predicating of it an infinite divisibility. To say that the atom is indivisible and impenetrable and yet extended is a self-contradiction. There are atoms, but they are not extended or parts of extension, for they are indivisible and without parts. These atoms are the monads. They are spiritual entities. They are the subjects of experience and their activity is perception. The universe is mirrored into each monad. To this doctrine Leibniz was led by considering the contradictions in the old concept of atoms and the void. I will quote his own account written in the last year of his life.

" When I was a youth I accepted the void and the atoms, but reason saved me from the imagination which makes sport of us. Imagination limits our researches, fastens our meditation as it were with a nail, makes us think we have found the ultimate first element, the *non plus ultra*. We want nature to stop where our imagination reaches its limit ; we want nature to be finite like our mind ; but this is to fail to rise to a knowledge of the greatness and majesty of the author of nature. The minutest corpuscle is actually subdivisible to infinity, and contains a world of new creatures which would be absent from the universe if the corpuscle were the atom, that is, a body all in one piece and with no subdivisions."

It is important to notice here that Leibniz conceives the monads to be in their number actually infinite whereas the atom was conceived as in its nature finite. It is interesting to see also how this concept of the nature of the monad is connected with the rejection of the other article of the old theory, the void. I have already referred to Leibniz's argument against the void as being derogatory to the perfection of God. Here is an allusion to that argument which I quote because it connects it with his general doctrine.

" Leaving aside other reasons against the void and atoms, here are those I base on the perfection of God and on the principle of sufficient reason. I assume that every perfection which God could put into things without derogation to the other perfections he has put there. Now imagine to yourself a space completely empty, surely God could put some matter there without derogating at all from other things. It follows then that he would do so, consequently there is no completely empty space : all is full. The same argument proves that there is no corpuscle which is not subdivisible."

Such was Leibniz's attitude to the old atomic theory. What then was his position with regard

to the doctrine of Descartes? From his earliest
period his interest and his research was turned
inward on the mind rather than outward on nature.
The main point of the Cartesian system which
exercised him was the theory of the relation of
mind and body. In that relation the irreconcilable
nature of the dualism of the two substances, each
distinguished by an essential and mutually exclu-
sive attribute, is most pronounced. The logical
development of Descartes's principle in the
monistic philosophy of Spinoza only served to
emphasize the poverty of the concept. Leibniz's
train of thought was probably along some such
line as this. It started with the concept of God
and creation not as an assumption introduced to
explain the existence and origin of the world, but
as presenting the problem of existence and origin
in its most complete and definite form. Give
God, matter and movement, could he then, as
Descartes supposed, create a world? Clearly
not, for what God has created are finite individuals
active subjects, moral agents. These individual
activities are the real existences, and the universe
is wholly composed of them. What we have to
study then is the nature of these active substances,
the monads, the real atoms of nature, their

relations to one another and how they come to form a world. Take then my own existence. I am a monad, an active centre, an agent, the whole universe is mirrored into that centre, focussed there, and my activity consists in perceiving. The whole universe consists of my perceptions, but only an infinitesimal portion of these are clear and distinct perceptions, the rest are massed together, confused, obscure and undiscerned. I am also self-conscious, aware of myself as perceiving. My monad, the monad which is me, is apperceptive. But then I am in relation to a body, my mind is a dominant monad, and it works in complete harmony with the body, and yet this body is totally different and distinct in its nature from the mind. What is it ? It consists of monads but of inferior monads. They are infinite in number, for no principle exists which imposes a limitation on them, and yet each is individual, an active centre mirroring the universe from its own point of view. In this relation of mind and body there is clearly a harmony, and it is an original harmony. It cannot have been brought about by chance for it is of the essence of the relation. Here then in this fact of mind and body I have a reality which is a compound, decomposable to infinity, and yet consisting of simple elements which are individual and therefore indivisible. But this is typical of the whole universe and reveals its nature and origin. The harmony is pre-established, for it is the essence of the reality. If the universe came into existence by an act of creation the harmony entered into the creative design and was brought into existence with the universe.

Such with its necessarily theistic form seems to me to be the train of thought which produced in Leibniz's mind the idealist principle which he proposed to substitute for the materialistic principle, condemned on the ground first of self-contradiction and secondly of failure to satisfy the principle of sufficient reason. Though all the monads are alike in their nature they differ in their degree. This difference of degree is wholly concerned with the ideas of the monad. Each monad is in a necessary relation to all parts of the universe for its ideas are relative to the whole universe. And further, the monads do not differ from one another in the number of their ideas, for this number is infinite, they differ only in the degree of clearness which their ideas possess. Accordingly Leibniz supposes a hierarchy of the monads based on the clearness or obscurity of their ideas, that is, the perceptions in which their activity wholly consists.   There are three classes of created monads : (1) Simple monads, the elements of matter which have no clear thought ; (2) the souls of animals, which have some clear but no distinct ideas ; (3) finite minds, which have confused and also some clear and distinct ideas.   The supreme or uncreated monad, God, has only adequate ideas.

The meaning is that, taking my own mind, for example, as the monad, the whole universe consists in its perceptions, there is no passing beyond perception, or as Leibniz said, there are no windows by means of which anything can come in or pass out, each mind contains the universe. But whereas I have certain clear ideas and certain distinct ideas, the great mass of my perceptions outside these are a confused, obscure, blended heap.   Just as, to take one of his illustrations, the sound of the waves on the seashore consists of an infinite number of small sounds, which are not heard by me as each clear and distinct, but in the blur of one agglomerated undifferentiated mass.   There are degrees therefore of confusion and obscurity ranging from its upper limit in the absolute adequacy of the perceptions of God, all of whose ideas are clear and distinct, and its lower limit in the indefinite, possibly complete, confusion of perceptions in the simplest monad.

The difference between Newton and Leibniz as to the nature of the elements of matter is brought out with clear decision by Voltaire.

"The opinion of Newton is perhaps as modest as human opinion can be. It is limited to believing that the elements of matter are material, that is, there is an extended and impenetrable existing thing into the inner nature of which we enter ; God can divide it to infinity or he can annihilate it : he has not done so, and he preserves its extended and indivisible parts as the basis of all the products of the universe.

" Perhaps on the other hand no bolder theory than that of Leibniz has ever been put forward. Setting out from the principle of sufficient reason he tries to penetrate even into the deepest origins and into the inexplicable nature of the elements. Each body, he says, is composed of extended parts, but of what are these extended parts composed ? They are actually divisible and divided to infinity, therefore you can never find anything which is not extension. Now to say that extension is the sufficient reason of extension is simply to argue

in a circle and affirm nothing at all.   The ground
or cause of extended beings must be in non-
extended beings, that is, in simple beings or
monads.   Matter is therefore an assemblage of
simple beings."

If I have succeeded in making the doctrine of
the monads understood, it will be seen that it
is quite as inconsistent with the concept of
absolute space or void as it is with the concept
that extension is the essence of material substance.
The monads being non-extensive but all-inclusive
activities, their relations, if we describe them as an
assemblage, cannot be juxta-positions in a space
external to them and indifferent to them.   The
universe mirrored into each active centre is not a
particular part of some vast expanse conceived as
a container.   Space therefore is a reality which
must fall within the universe as the monad per-
ceives it mirrored, and it cannot fall outside the
monad, even were the concept of outside consistent with the affirmation of the monad.   For
Leibniz therefore space is neither a clear nor a
distinct idea, and consequently space does not
exist for God.   It belongs to our confused and
obscure perception, and in fact denotes the
obscurity and confusion which is inherent in

our view of the universe.   It is the mode in which
we present to our mind as co-existent with our
clear and distinct ideas the infinite residue of
indistinct perceptions.   Space therefore is not
a thing, not something which is, it is an order.
In this respect it holds precisely the same rank
as time.   Space is the order of co-existences,
time the order of succession.   Neither is a reality,
both are names for the confused blur of perceptions against which, as against a background,
our clear and distinct ideas stand out.

There is then a complete contrast between the
two conceptions—the idealist universe of Leibniz,
the materialist universe of Newton.   The ultimate
principle of the first is universal mind, represented
as a supreme God, perfect in wisdom and infinite
in power.   His need of creation is not a mechanical need, founded in some impulse, rational or
irrational, to construct complex systems or direct
simple movements to complex effects, on the contrary it is purely a spiritual need, and creation is
the bringing into existence of active subjects, so
perfectly harmonized in their range, so fitted into
the scheme by their degree, that no one is redun-
dant and no possibility unrealized.   God can
create or annihilate the universe, but not in part,

H

only as a whole. This was the famous concept of " the best of all possible worlds," which since Voltaire has mainly served as a theme for jest. Divested of its theistic setting, and studied as it should be, that is, as a metaphysical research into the nature of reality and the origin of our concept of external nature, it will answer any logical challenge.

Newton's conception on the contrary has commended itself to the scientific mind, and it has in its pre-suppositions unquestionable advantages in practice. Logically and metaphysically it is riddled with contradictions, and these appear most strikingly in its theological consequences. Newton could not be indifferent to the theistic difficulty, however little importance it may seem to have had for his successors. Hence the theory of the sensorium. It was in the last years of Leibniz's life (1715-16) and in Newton's old age that this theistic problem was discussed in the correspondence between Leibniz and Clarke, one of the most valuable of the philosophical remains included in the editions of Leibniz's works. It was an unfinished correspondence, for it was interrupted by death.

The occasion of this correspondence was a

letter which Leibniz wrote to the Princess of Wales, afterwards Queen Caroline, wife of George II., in 1715. In it he said : " It seems that even in England natural religion has grown very weak. There are many who hold that souls are corporeal and others who even hold that God himself is corporeal. Newton says that space is the organ which God uses in order to be conscious of things. But if God is in need of means in order to be conscious of things, it follows that these things cannot wholly depend on God. They cannot be his production. Newton and his followers have a still odder notion of God's work, for according to them God has every now and again to wind up his work as we wind up a watch which would otherwise stop. God has not, it seems, had enough foresight to give his work perpetual motion. Indeed the machine which God has made is so imperfect, according to them, that it requires polishing up every now and again by a special effort, and even needs regulating. Like a watchmaker he reveals the defects of his watch by the number of times he has to correct and retouch it. In my view the same force and vigour is everywhere in evidence, passing from one thing to another according to laws of nature and the perfect order pre-established.   If God performs miracles it is not because nature requires them, it is on account of grace.   To judge otherwise is to entertain a very low idea of God's wisdom and power."

The sting of this letter so far as it concerns Newton is the reference to the theory of the sensorium.   The Princess asked Newton to reply to it and he commissioned his disciple Clarke to defend him.   It is in the correspondence which followed that Leibniz expounds with great clearness his theory of space.   Real absolute space, he declares, is the idol of some modern English. He uses the word idol, he explains, not in its theological but in its philosophical meaning, and he quotes Lord Bacon's *idola tribûs, idola pecûs.* If space is a real thing, as these writers contend, then it is eternal and infinite, and they must identify it with God.   Either it is God himself or it is an attribute of God, his immensity.   But then space has parts—how does that apply to God ?   " Space for me," he says, " is purely relative as also time is.   Space is an order of co-existences as time is an order of successions. Space marks in terms of possibility an order of things which exist in the same time so that they exist together. The manner of their existing is not in question." When we see several things together we perceive the order in which they stand to one another. Were it otherwise there would be no sufficient reason for the world being here not there, now not then.

In Clarke's reply there occurs a remarkable passage intended as a *reductio ad absurdum*, but which sounds to us almost as an anticipation of the negative result of the Michelson-Morley experiment which led to the first theory of relativity. " If space were no more than the order of co-existent things, it would follow that should God make the whole world move in a straight line with any velocity he liked to choose, it would still be always in the same place, and when the movement ceased nothing would sustain a shock."

We have then in the contrast between the principles of Newton and Leibniz, the distinction, in its full intensity and most emphatic expression, between a nature-philosophy based on a materialistic principle and a nature-philosophy based on a spiritualistic or idealistic principle. Both principles are directed to mathematical and physical researches, both are conceived as the true principles which underlie the science of nature.

Neither has stopped dead, neither has ceased to undergo development, but each has chosen a separate and a widely divergent path. So far as experimental science is concerned the principle of Leibniz has been rejected absolutely, at times with contumely and scorn, while Newton's principle has seemed to be confirmed by the sure advance of science during the last two hundred years. It is with a shock of surprise that we are receiving to-day a challenge to that principle which seems to question the very foundations of the concept on which it is based. Still more strange is it that physical science itself is seeking a principle which will enable it to co-ordinate observations from individual centres of experience (monads), without the aid of, and recognizing the impossibility of having, absolute standards of reference independent of the observers. In effect we are proposing in mathematical and physical science to abandon Newton's philosophy and adopt that of Leibniz.

CHAPTER VII

THE MODERN SCIENTIFIC REVOLUTION
AND ITS LEADERS

IT is curious that we should associate the maxim *hypotheses non fingo* with the scientific method of experiment, and suppose that the making of hypotheses is a particular vice of speculative metaphysics. The exact contrary is true. It is the scientific experimenter who makes hypotheses, and it is no reproach to his method that he does so, it is its essential feature and by no other means can he advance. On the other hand it is the philosopher who vitiates his method the moment he proceeds to make hypotheses, for the assumptions he introduces into his premises will assuredly re-appear in his conclusions. It is the philosophers, of whom we may take Descartes and Hume as types, who are entitled to inscribe on their banners Newton's often quoted motto.

This does not mean that Newton was vainly boasting. He was indeed a true philosopher in his simple and direct acceptance of fact, his disregard of its conflict with accepted theory, and his attempt to interpret it without any respect for preconceived opinion. I shall endeavour to show that Einstein, in his formulation of a new theory of gravitation, in this respect follows his great predecessor, for the new theory is not a hypothesis, it is the acceptance of paradoxical facts in spite of the paradox, because they are based on experience and confirmed by experiment, and the formation of a principle in conformity with them.

It will be well to begin by stating in definite and positive terms what the new principle is without reference to the experiments which have called for it.

The relativity principle of the classical mechanics supposes an absolute space, an absolute time and an infinite, that is, an infinitely variable, velocity. Transformations of measurements for different systems of reference, spatial or temporal, can be simply and arithmetically calculated and expressed as variations of velocity. Euclid's postulates are applicable and universally valid. Observational deformations of Euclidean figures are apparent, not real, and easily explicable as perspectives, due to the conditions of the observer. Let me illustrate it by supposing that I travel, say from London to Edinburgh, in a slow train while you travel the same route by an express,—the space and the time have not varied, they are identical for each of us, but the apparent stretching of the space and time for me, and their apparent contraction for you, are simply adjusted by taking into account the velocity of the train, say thirty miles an hour for me, sixty miles an hour for you. We express this relativity by saying that we have moved through the same space, you in half the time lived through by me. This is the commonly accepted principle.

The new principle of relativity has two stages, the special principle and the general principle. The difference is not intensive but extensive. The first was confined to a definite phenomenon, the velocity of the propagation of light *in vacuo*. The second is the application of the same principle to all laws of nature. The special relativity principle is based on the fundamental concept that space is variable and that time is variable, and that there is one finite velocity which is constant and also a limiting velocity, that is, a velocity which cannot be exceeded, but to which the approach is infinite. This constant velocity is that of the propagation of light. The concept of this velocity as constant and as a limit is not as arbitrary as it at first appears to the uninitiated, for as matter of fact we are absolutely dependent on light signals for sending messages, or measuring distances, or estimating intervals between events. The special principle of relativity is that observers in systems moving uniformly in relation to one another do not use the same standards in measuring space and time dimensions, but each observer employs a standard which supposes his own system at rest. When then there are observers of the same events, in different systems of reference, moving uniformly relatively to one another (as in the two trains), each observer sees the light signals propagated with the same velocity, but the difference due to the movements of the two systems, relatively to one another, is in the space and time which are different. It is easy to illustrate. Let us go back to the two railway journeys. According to the classical mechanics one is double the velocity of the other. According to the principle of relativity the velocity of each is identical because in each train the observer is at rest. The difference is in the space and the time. These are elongated for the traveller in the slow train, shortened for the traveller in the express. To common-sense this appears contradictory, but reflection will show that it is a simple alternative to the common-sense view and logically an exact equivalent. It is simply equal to saying, what is also fact, that in our two journeys neither I nor you moved at all, but our destination moved to us, and in doing so traversed double the space in double the time in coming to me that it did in coming to you.

The special relativity principle was confined to the consideration of observers in systems of reference in *uniform* movement of translation relatively to one another as, for example, in the two trains. The general relativity principle is the extension of the special principle to non-uniform systems, in particular to rotational systems, and it affects the concept of the laws of nature. It declares that when the movement of a system relatively to other systems is non-uniform, the peculiar quality which the observer experiences as influence or force can find an equivalent expression in the motion of the system for an observer in another system. The rotational movement of a system which gives rise to the phenomenon which an observer at rest within it describes as weight or gravitation, that is, the falling of loose bodies towards the centre of rotation, might appear to an observer in another system to be simply the movement of the rotating system itself towards other systems moving non-uniformly in relation to it. It further deduces that for all non-uniform systems the spatial coordinates (length, breadth, depth) are not Euclidean. In a gravitational field space for all observers is warped or curved. For the moment I am trying to state what the principle is without explaining the grounds for adopting it. Thus, to return to the illustration of the train journeys, in the two previous uses of the illustration I supposed the movement uniform, and there was consequently little difficulty in showing that it is exactly equivalent to regard the train as not moving and the destination moving, or to regard the train as moving and the destination as fixed. In the one case we shall say that the space and time vary and the velocity is constant, in the other that space and time do not change but velocity does. But now suppose that the train is brought to a stop by the sudden application of the brakes and that the traveller is thrown heavily from his seat. There will be difficulty in making him take the view that he was all the while at rest, that it was the seat which moved from him and the floor which moved towards him. Yet this is exactly how it might appear to an observer in another system, and in so describing it he would have an equal right to declare it true.

Suppose then we understand the principle and accept it to the extent of admitting that we can, if it is worth our while, always find an equivalent way of describing the phenomena of movement in place of the one to which we are accustomed. Is it then, we shall ask, merely a matter of choice? If not, what is the necessity for disturbing our ingrained common-sense principles? The reply is not an argument, but an experiment.

The experiment which has occasioned the revolution in our fundamental concepts and called for the formulation of a new basis of a philosophy of nature was carried out by Michelson and Morley in 1887 and is described in the *Philosophical Magazine* for December of that year. It has been so often described since that it will be

sufficient to allude to it here and go straightway
to the principle of it.   Since Newton, and on the
foundations he laid down, there has been a steady
and continuous evolution of scientific knowledge,
marked by an ever widening range, an ever
deepening insight, and an ever surer inclusiveness.
The theoretical objection to his theory of gravitation that the concept of action at a distance
which it involved is a concept of something
wholly unimaginable, was made by Leibniz.
Newton himself felt the force of the objection,
and to meet it supposed that space might be filled
with a subtile etherial matter.   The theory of the
ether of space became a necessary hypothesis,
however, only when the undulatory theory of light
was formulated.   With the extension of the discoveries of electro-magnetic phenomena this ether
tended to become one of the fundamental concepts
of science.   It was, however, purely conceptual,
and the famous Michelson-Morley experiment
was contrived with the primary intention of testing
its reality as a physical existence.   The principle
is easily explained.   The earth is in a movement of
translation relatively to the sun amounting to about
30 kilometres a second.   If then we consider the
earth as at rest we must suppose the ether to be

streaming past it at the equivalent velocity. Now
it is easy to calculate the difference in time
required for any given uniform propagation to
reach a point and return in a direction which
crosses the stream, and to reach and return from
an equally distant point in the direction of the
stream.   Thus, let us suppose, for instance, that
the stream is flowing 30 kilometres a second, and
that our signal unretarded would travel 50 kilometres a second, and that we project it to a
measured point in each direction, up the stream
and across it, from which points it is immediately
reflected and returns to us.   Across the stream
the flow will be equal to a retardation of 10 kilometres a second, our projectile will therefore
travel 40 kilometres in a second and will return
in the same time, that is, it will go and return in
two seconds.   Now suppose it is directed up
stream, it will have the full retardation and therefore will travel only 20 kilometres in the first
second and to arrive at the measured point, 40
kilometres, it will require two seconds, the same
time in which, across the stream, it went that
distance and back.   It is true it will return in half
a second, for it will have the 30 kilometres a
second added to its own velocity of 50, but it will

be that half-second later than the other.   In the actual experiment a beam of light was sent to a mirror and reflected back along each axis.   Light has a velocity approaching 300,000 kilometres a second, but the calculation was worked out to an accuracy, and within a margin of error, which made it still quite certain that the retardation would be revealed.   To the great surprise and deep disappointment of the experimenters the result was negative.   Along each axis the return was simultaneous.

This was the famous experiment ; let us now look at its consequences.   In the first place it negatived the ether hypothesis.   There is no ether if by ether is meant something occupying space and at rest relatively to the matter which moves through it.   If there be ether it must move with the system which is in relative translation (exactly as Descartes supposed).   If to suppose an ether is to suppose infinite ethers, then the hypothesis is useless and may be dismissed.

A second consequence is that we must suppose the velocity of light *in vacuo* to be unaffected by the movement of the source.   This is more fundamental still, for if there be no ether or if the ether be perfectly frictionless, the velocity,

when the source is moving, ought to be different
from the velocity when the source is at rest, yet
the experiment proved it to be the same in each
case.

A third consequence is that as the laws of
nature remain constant for observers in moving
systems, the space and time dimensions in order
to keep the constant ratio must themselves vary. $\curlyvee$

A hypothesis to account for the negative result
of the experiment was put forward independently
by the late G. F. Fitzgerald, Professor in Trinity
College, Dublin, and H. A. Lorentz, Professor
of Physics in the University of Leiden. It is
generally known as the Lorentz-Fitzgerald contraction. It assumed that the dimensions of a
solid body moving through the ether undergo
slight changes, that the moving body is automatically contracted in the line of the direction in
which it is moving, and that this contraction is
exactly equivalent to and counterbalances the
difference which would otherwise be manifested
in the velocity of light. It will be seen that such
hypothesis is primarily nothing but a simple *ad
hoc* device for giving expression to the negative
result of the experiment, and is little more than
an acknowledgment and way of stating positively

I

the fact observed.   It has, however, been possible to submit the question whether there is in reality, or as physical fact, such a contraction, to laboratory test, by measuring the electrical resistance of metal rods both across and in the direction of their translation.   These experiments have proved negative and have caused the abandonment of the hypothesis by physicists, in so far at least as it purports to be the actual explanation of the phenomenon.

A much more radical interpretation was given by Hermann Minkowski in 1908, shortly before his lamented death at the age of forty-five, in an address on " Space and Time," which has since become historical.   In it he formulated a new mathematical theory which embodied in a complete form the principle, the necessity of which had been revealed in the negative result of the experiment.   Minkowski proposed a new mathematical scheme of the universe in which time entered as a dimension.   The universe is to be conceived not, as hitherto, as a three-dimensional continuum enduring in a one-dimensional time, to which it is indifferent, but as a four-dimensional continuum in which the three dimensions of space and the dimension of time are the axes of coordination by which the observers in systems of reference moving relatively to one another relate the constituent factors of the universe, that is, the events the assemblage of which is the universe. The beauty of this theory is that its apparent strangeness when first propounded tends to give way to familiarity and obviousness ; for when we come to think of it we recognize that the world of our living experience is four-dimensional. The opening sentences of the address (delivered at Cologne on September 21, 1908) have been often quoted as marking the beginning of a new epoch in physical theory. " The observations concerning space and time which I am about to expound are the results of experiments in physics. Therein lies their strength. They go to the root of matters. Henceforward space and time as independent things must sink to mere shadows and the only thing which can preserve some sort of subsistence is a kind of union of the two." He then developed his argument. The universe must be conceived as an assemblage of events. An event is determined for every observer from the standpoint of his system of reference and in relation to that system, by four coordinates, three for space and one for time. Suppose two events to occur, observers in one system of reference will measure a certain distance separating the two events in space and a certain interval separating them in time. These measurements will not accord with those made by observers in other systems of reference who will measure the same events from their own standpoint and will find different distances and different intervals. Each observer will keep constant the ratio between his co-ordinates of measurement in passing from one system to another, but the four axes themselves will each undergo variation.

The simplicity of Minkowski's scheme won for it general admiration and made the adoption of the principle of relativity easy in mathematics. A fourth dimension of space had hitherto been always associated with the attempt to rationalize the claims of spiritualistic phenomena to actuality, but Minkowski showed that the fourth dimension was necessary to explain experience on the ordinary common-sense plane. Einstein, referring to this four dimensional theory of Minkowski, writes : " A mystical shudder seizes the non-mathematician when a fourth dimension is spoken of, a creepy emotion, something like that produced by a stage effect. Yet there is no more obvious commonplace than that our world of everyday experience is a four dimensional space-time continuum."

There was, however, a deduction of the experiment which presented the aspect of complete paradox. This was the constancy of the velocity of light for all observers and its independence of all variations in the relative velocity of a system. The velocity of light *in vacuo* is 300,000 kilometres a second, it is a finite velocity, and no velocity is known to exceed it. Some of the radio-active substances give off particles which approach it—the β particles—but though they attain to 99 per cent. of the velocity they are not known to reach it. Now suppose that relatively to ourselves there is a system being translated at the rate of 150,000 kilometres a second,—for observers in that system as for us light is propagated at the uniform finite velocity of 300,000 kilometres a second. This appears a direct self-contradiction. How is it to be reconciled? However surprising it may be to have to acknowledge that we can find no evidence of the absolute movement of a material system nor discover the direction of that movement by experiments made within the system, it is more than surprising, for it seems downright absurd, to affirm that a finite velocity is identical for all observers in all systems of translation, uniform or non-uniform in relation to one another. It is in regard to this problem that the work of Einstein is particularly important. His philosophical attitude towards this paradox is specially deserving of notice.

Einstein from the first accepted the negative result of the experiment. It did not for him indicate an agnostic position. It did not merely mean that an absolute movement, or movement measured from an absolute or fixed zero, is unknowable. He accepted the negative evidence as definite proof of the non-existence of an absolute standard, and he proposed to reject the postulate that an absolute standard is a necessity of thought. But how is the rejection of absolute movement compatible with the affirmation of a constant finite velocity? According to Einstein this incompatibility is purely apparent, in fact it is compensated by the deformation of the axes of co-ordination used by one observer as seen by another. Thus to an observer in a system moving relatively and uniformly to us at half the speed of light our proportions are foreshortened to half what

they appear to us, so that measuring the propagation of light our unit is double that of his, and
his is correspondingly half that of ours. Each
observer therefore finds the light propagated
at the same velocity of 300,000 kilometres a
second, but the kilometres used by the one appear
to the observer in the rapidly moving system
elongated to double their length, and those used
by the observer in the rapidly moving system
appear halved in their proportion to the observer
in the slow moving system. \

The special principle of relativity which Einstein formulated in 1905 had in view this problem
but only in regard to the velocity of light and its
independence of the movement of the source.
Already, however, it had seemed to him that if
this principle be true it is not limited in its
application and it probably implies an entirely
new conception of physical reality. The general
principle of relativity, which formulates a new law
of gravitation in place of the law of Newton, is the
result. It is not based on a new experiment but
on an application of the principle of the original
experiment to all the laws of nature. In 1917
Einstein had thought out the three means by
which his new theory could be brought to the

test. First that it would account for the discordance of the motion of the perihelion of
Mercury, as calculated on the Newtonian formula
and as it is found to be in fact. Its cause had
been sought for in vain in the supposed presence of a mass of matter between the orbit
and the sun. It had also been suggested that
the Newtonian law may require an infinitesimal alteration in its mathematical formula to
make it bring out the result correctly. But the
idea of the existence of an unknown minor planet
is now generally agreed to be so improbable as
almost to amount to the certainty that there is
none ; and an alteration of the mathematical
formula to make the calculation of Mercury's
period come right would make other calculations
work out wrong. In this case Einstein's theory
based on the principle of Relativity has been so
triumphantly vindicated that to many mathematicians and physicists it is sufficient of itself to
establish the principle. Einstein's formula gave
the exact result without upsetting the calculations
in any other case. The second means was the
displacement of stars in the gravitational field.
He predicted that the light of stars near the sun
would be deflected and that the deflection would

be twice the amount which Newton's law, calculated for the mass and velocity, would give, if light, as Newton thought possible, had mass. This prediction was verified in the observations of the Eclipse Expedition of May 1919, and was the occasion of the extraordinary public interest in the new theory. The third means proposed has so far not passed the test, and is the subject of research and of much discussion. The vibration of atoms on the surface of the sun compared with the vibrations of atoms on the earth ought, Einstein says, by reason of the difference of the gravitational field, to show a shifting of spectral lines towards the red end of the spectrum.

I will now try and explain by illustrations what the new principle is. Let us suppose, then, that from the window of a smoothly travelling railway carriage a stone is dropped, and that an observer in the train watches its fall. For him it follows a straight course to the ground. An observer watches the same event from a position which is fixed in respect of the moving system of the train. For him it follows a curve. Which is right? Is there a real straight line which is the shortest distance between two points? Is there any way by which each observer can so correct his observation that he can discover the real as distinct from the apparent line ? The principle of relativity declares that there is no way of deciding between the two observers, that each must use the coordinates he carries with him and that these adapt themselves to accord with every system of reference he enters, and that each observer therefore measures an event from the standpoint of his own system regarded as at rest. From this it follows that an event (dropping the stone from the carriage window) which for one observer occurs in one and the same place, for another observer has its beginning separated from its end by a distance in space. If we consider the whole event as single, then the point at which it is observed from one system of reference will not correspond with its place as observed from another. This is the principle of relativity as it applies to space.

Now suppose that time not space is in question, and that observers take positions equidistant from the point where an event is to take place for the purpose of recording it simultaneously. Let two observers be placed at an equal distance from a point on an electric railway at which it is arranged that a break in the circuit shall be indicated by a flash. Each observer will see the flash at the same instant. Suppose now that two other observers are similarly situated, but on the moving train instead of on the permanent way. It is true that in such case, as the timing is with light signals, it will not be possible to make any difference appreciable. The velocity of light is so disproportionate to any conceivable velocity of the train that the effect due to that difference could only be expressed in thousands of millionths of a second in time or of an inch in space, but, all the same, the principle can be made clear by the illustration. If the precise moment of the emission of the signal (the electric spark) is the same for two observers equidistant on the permanent way, it cannot be the same for two observers equidistant in the moving train, because during the propagation one observer will have advanced to meet the signal, the other will have receded, giving it a correspondingly longer route to travel. The principle of relativity is therefore that simultaneity has no absolute meaning. No two events which happen at the same moment for observers on one system of reference can be simultaneous for observers in another system. Therefore, two events which for one observer are simultaneous, for another are before and after.

So far we have been following the special principle of relativity. It teaches us that, contrary to the notion of classical mechanics, according to which all differences in the observations of events, and all difference in the appearance of events to observers, are calculable in terms of a constant space and time and a variable velocity, there is no distance in space and no interval of time which is invariable and independent of the system of relative movement to which the observer is attached. It takes us a long way in the direction of a new coordination of nature on a philosophical principle, but there is nothing in it peculiarly subversive of our ordinary concepts. Indeed to many mathematicians it commended itself at once as seeming to offer a much better scheme on which to undertake the organization of the science of nature. When, however, the principle is applied not only to systems moving uniformly in relation to one another, and to phenomena inappreciable in ordinary experience, such as electro-magnetic propagations, but also to gravitation and the ordinary laws of nature, it touches the science and mathematics of common life. It is then that it disturbs our feeling of at-home-ness in the universe, brings over us a feeling of giddiness and makes it seem impossible at once to attain a new equilibrium.

There is something fundamental in our experience of weight. Without it the world would seem to have no substance. Without it we should feel like the man in the folk tale who was induced to barter his shadow. Now it is easy enough to imagine that the phenomena of gravitation may be unknown to observers in other systems, but to suppose that they may observe the identical phenomenon which we experience as weight and yet observe it not as weight but as the movement of the system, and that this movement is the exact equivalent of what we experience as force—this is very difficult to accept. It is precisely this that the general principle of relativity affirms.

Suppose, to take one of Einstein's illustrations, a room, such as that in which the reader may be sitting, to be detached and transported bodily into some distant region so remote from masses of matter that it is entirely free from any force of attraction. Let us suppose no gravitational field. We shall have to think of it as artificially held together and the objects in it as fastened by cords or such like means, for the loose objects it might contain would follow the direction of any chance movement which might be imposed on them. If we suppose ourselves still sitting or standing on the floor it will not be by reason of our weight but by attachment or holding on. Let us now imagine that this room is attached by a hook to a cord outside which is being pulled, thereby drawing the room in a definite direction. The hook being attached to the ceiling that definite direction of the ceiling will be upwards. At once all loose objects will lie and remain lying on the floor, and free suspended objects will hang downwards, and we, if we would rise from the floor, will have to put forth an effort sufficient to produce a movement in excess of the dragging movement. To us in the room it will seem then that objects have suddenly become heavy and that they fall downwards by their weight, but to observers outside the whole phenomenon will be a simple consequence of the dragging movement and explicable in terms of it. If an object is detached it will not, as seen by the outside observer, fall on the floor, it will remain immobile till the floor reaches it. This is what Einstein means by equivalence. What appears to the observer in a

rotating system as weight, that is as the attraction
of the object to the centre of the rotating mass,
will appear to an observer in another system as
the movement of the rotating mass to the object
at rest.

Very curious consequences follow, completely
subversive of our ordinary ideas. Gravitation is
a phenomenon which is connected with a rotational system, and the gravitational field is the
space which surrounds this rotational system.
To an observer attached to the system, regarding
the firmament from the standpoint of his system at
rest, the firmament is in movement. Consequently
for this observer an object (Newton's apple)
detached from his system moves with the firmament. But to an observer on another system at
rest, for whom the first system is rotating, the
detached object (the apple) ceases to move with
the rotating system and remains at rest. Therefore to the first observer the movement of the apple
will be its fall towards the centre of the rotating
earth, but to the second it will be the movement
of the earth towards the apple. The gravitational
field is then the moving firmament, as observed
from the rotating system, by an observer on the
system taking the standpoint of himself at rest.

It is clear that the nearer the object in the field is to the rotating system the greater is the velocity of its observed movement. What we call an attractive force decreasing inversely with the distance is the exact equivalent of the movement of bodies in a gravitational field, as observed from a standpoint of a rotational system at rest.

This enables us to see what is meant when it is affirmed that space in the gravitational field is non-Euclidean. In the gravitational field the observer is either at rest in the firmament relatively to the rotational system or else at rest on the rotational system relatively to the moving firmament. In neither case can Euclid's postulates be fulfilled or Euclid's axioms hold true. Consider, for example, the postulate that a straight line is the shortest distance between two points. The shortest distance for one observer will not be the shortest distance for the other, or, what is the same thing, the straight line drawn by the one will appear curved to the other. But when we realize this relativity can we not make allowance for the appearance to different observers and work out the perspective law for each ? This is what we imagine we are always engaged in doing.

Is it not a commonplace of psychology that visual space is not real space? The principle of relativity shows us that this supposed power of correcting appearance by reference to an intellectual absolute is illusion.

The illusion is persistent. It is so easy to represent our system of reference in movement by means of our imagination, that we have come to cease to think of the firmament as in movement and, though it still appears so, we translate the appearance automatically into the reality of our own movement. Can anyone really think the smoothly running railway carriage is at rest while the landscape is in movement? We fail therefore to see that though our imagination aids us in representing our own movement as relative it never brings us to a system at absolute rest. That necessary standpoint of a system at rest must be within the observer himself and is the condition of his observation. It is from that standpoint that our measurements are made, and we carry with us, attached to us, inseparable from us, the axes of coordination which we apply to the universe. We are therefore perpetually under the illusion that the absolute criterion lies without us, whereas it is indissolubly part of our nature.

It is just as if we supposed the centre of a circle to be independent of its circumference and free to move about within it, even indeed to come outside and survey it.

Common experience offers us an example of a persistent illusion in regard to space. We think that we know by actual experience real space. But the space we think real is not the space we perceive by any of our senses. We do not think, for example, that the sun and the moon are in reality the same size as a threepenny bit, though they cover the same amount of visual extension as this object held at arm's length. The space we think real is purely conceptual and ideal, and this space is Euclidean. We do not encounter it in experience, nevertheless we are convinced it really exists and suppose it must underlie and condition our experience, and we distinguish it as noumenal from the phenomenal. It is this persistent illusion of a real space, noumenal not phenomenal, which makes it so disconcerting to us and so difficult to accept the notion of a space which may be curved, warped or distorted, not in its appearance only, but in reality. It is affirmation of a space with such properties that makes the general principle of relativity appear paradoxical.

If, however, we accept the negative result of the physical experiments we are bound to reject as pure illusion the notion that Euclidean space exists beneath the world of sense perception.

We may make the concept of curvature in space clear by simplifying for the purpose of illustration the scheme of the principle. Let us suppose then that instead of manifold gravitational fields there exists one only. Let us abstract, that is to say, from everything in nature except the rotating earth and the surrounding extension relative to it which we will call the firmament. The gravitational field will now consist of two and only two systems of reference. An observer at rest on the earth sees the firmament in a continuous movement from east to west. An observer in the firmament sees the earth rotating on its axis from west to east. The surface of the earth will then be the centre of a gravitational field, and the limits of this field will be the earth's axis on the one side, and an external imaginary limit of a space purely Euclidean on the other. It is clear that within this gravitational field space must be warped, because every observer will be at rest on one of the two systems. Any movement, therefore, be it of propagation or of translation, which takes place in the gravitational field must be occurring in the space of one of the systems.   Suppose then any object to be detached from the rotating earth, it at once takes its place in the circulating firmament.   Similarly, any object detached from the firmament takes its place in the rotating earth. The universal space simply denotes the continuity of these two systems in the relation of movement and rest for one another.   Space is not a third somewhat in which these movements are taking place.   To affirm it would simply be to deny the reality of one of the movements, and what possible ground can there be for that ?   The essence of the principle of relativity is the explicit denial of such an absolute space and the recognition of the impossibility of interpreting facts on the supposition of its existence.   The curvature of space is therefore physical fact, not subjective appearance.   Now let us suppose a straight line drawn from within the earth system to any point in the firmament system, the course of that straight line must *by its spatial condition* curve in its path through the earth system and curve in the reverse direction when it passes through the firmament system.   Any line which is the shortest between two points for an observer in one system will not be the shortest for an observer in the other.

Someone may still object. He may demur that my conclusion that there is real curvature of space in the gravitational field follows only because I have started with an arbitrary hypothesis, viz. that the universe is a single gravitational field. The reply is that the conclusion does not depend on this hypothesis and it is only introduced for simplicity and to enable attention to be directed on the essential point. Admit that the universe is infinitely complicated, that it consists of infinite systems moving uniformly and non-uniformly relatively to one another, this only increases the difficulty of correlating observations. No doubt this difficulty has been the guiding motive in the evolution of the concept of absolute space, but it is not a proof that absolute space exists.

Three hundred years ago it was discovered that a heavy body such as lead and a light body such as wood follow the same identical course in the same identical time in the gravitational field, when allowed to fall with the same initial velocity and with all frictional obstructions removed, as for example in the vacuum produced by an air pump. The significance of this discovery was never thought out. It is precisely what is affirmed by the theory of equivalence. Why does a stone fall to the ground when let go? The usual answer is because it was lifted off the ground, and the hypothesis in that answer is that the earth exercises a direct influence on the stone, called the force of attraction. The strength of this force was found to vary with the distance according to the well-known Newtonian law. As an explanation this has not satisfied modern physics because it admits the notion of action at a distance. The advance in the study of electro-magnetic phenomena has led to the conclusion that there is no action at a distance. When we see a magnet attract a piece of iron, science will not let us be satisfied with the simple explanation that the magnet draws the iron across empty space, science requires us to see in the phenomenon a property or character of the intervening space which it names the magnetic field. It is this and not the magnet which influences the iron, causing it to move toward the magnet. In gravitation we have the exact analogy and at the same time an important contrast. The contrast is that in the gravitational field (as Galileo's experiments proved), bodies undergo an acceleration which does not depend on their material or constitution. And in this way the principle of relativity explains gravitation, showing it to be a phenomenon which depends on relative systems of movement, and on the position or standpoint of rest, necessarily assumed by the observer for his own system.

What kind of world is it then that we live in? A world which is finite in so far as the concepts of space and time determine it, and which yet is not circumscribed. The principle of relativity enables us without doing violence to the laws of thought and without contradicting experience to dispute the Euclidean axioms. Space and time, which throughout the whole history of philosophy have been the stubborn realities of the framework of the universe, baffling the mind in every effort to form a consistent scheme of nature, are deposed, at least from their dominating position in the mathematical and physical sciences. It is a triumph of philosophy, for the principle of relativity is a return to the concept which Leibniz indicated and which was abandoned by the scientific successors of Newton. In nothing is the contrast more striking than in the concept of space. For Newton space is an infinite, absolute, immensity, which can only be present to the perception of an infinite God. For Leibniz God alone of intelligent beings is wholly without the perception of space, because God is conceived as an intelligence with adequate ideas and with no obscure perception. Let us drop the theological expression and state the same contrast in scientific terms. We shall then say that for the materialist space is the fundamental reality and the universe presupposes it ; for the relativist, on the other hand, space is a limitation of the observer's apprehension of the universe. Infinity is not the affirmation of space but its disappearance.

CHAPTER VIII

CONCLUSION: IN WHAT SENSE IS THE UNIVERSE INFINITE?

EVERY revolution in the world-view has profoundly affected mankind in those aspects of life which depend upon reason. So far as most of us are concerned the principle of relativity may seem a matter of small importance, dealing with infinitesimals which in the ordinary business of life are entirely inappreciable. It disturbs our general scientific methods no more than the Copernican theory disturbed the practical adjustments of the human mind. For mankind the sun continues to rise and set. We reckon the times and the seasons, as men have always done, and will do, irrespective of any change which has taken place, or which may take place, in astronomical theory. Newton's law of the inverse square will not cease to be a practical rule for engineers and mechanicians for all economic projects, nor will it cease to commend itself by its simplicity, if Einstein's formula comes to be recognized as theoretically perfect. In religion, however, and in philosophy of life—philosophy as it concerns mankind generally and not as technical metaphysics or theory of knowledge —its effect will be profound and far-reaching. I will conclude therefore with an indication of some of these higher interests in the new principle.

It seems to me that the new world-view must take the form of, and find the imagery for, a new concept of the nature of the continuity and infinity of the universe.   In the world-view as it has found expression in religion and philosophy hitherto the concept of infinity has been inseparably associated with the ideas of space and time.   " As it was in the beginning, is now, and ever shall be, world without end," is the liturgical expression of this idea.   It depends on the notion of the absoluteness of the spatio-temporal order.   The mathematical definition of continuity and infinity is, as we have seen, simply a precise form of expression for the spatial and temporal concepts, depends upon them for its applicability, and appeals exclusively to the pragmatic test.   It is justified because it works, but it only works in so far as we accept in advance the postulate of an external world in space and time. The new principle of relativity goes behind and beneath the mathematical definition of infinity, for it rejects the postulate on which it is based. So far as the mathematical principle rests on the Euclidean postulates and so far as infinity means, when applied to Euclidean space and its imagery, boundlessness and absence of limits, we have seen that the new principle definitely rejects the concept of infinity. It gives us in fact what to common-sense is a new paradox—a world which is finite and yet not circumscribed. So far, however, we have to do only with the negative aspect of the principle. What has it to offer on the positive side ?

The answer to me seems clear and manifest. We are offered in place of the contradictory pseudo-concepts of endless extension and limitless duration the concept of a truly infinite universe. The infinity of the universe is based on the nature of life and consciousness. The principle of relativity declares that there is no absolute magnitude, that there exists nothing whatever which can claim to be great or small in its own nature, also there is no absolute duration, nothing whatever which in its own nature is short or long. I coordinate my universe from my own standpoint of rest in a system of reference in relation to which all else is moving. That system may change, and there is no limit to the change it may undergo, but however great the change, measured by its relation to other systems, its dimensions remain constant. I, the observer, am not a point at an instant. Space and time dimensions do not apply to mind. A monad has no dimensions, so that one mind or monad can be in a relation of magnitude to another ; one monad does not occupy more or less space than another. Space and time are not containers, nor are they contents, they are variants. Consequently, whatever my system of reference, as I pass into it or out of it, that is, as it changes, so my spatial points and temporal instants change ; my units of measurement vary, so keeping the dimensions of my universe constant.

I may illustrate my meaning if I now give with as much imagery as the concept will admit my idea of the nature of the infinity and continuity of the universe. In doing so I will set aside all questions of detail and all special problems in

order to set out the scheme with as much genera-
lity as possible.

I start, then, from the monadic concept. We
all belong to the order of self-conscious minds.
Everyone has his own system of reference within
which from his standpoint of rest therein he may
correlate events which happen for him with events
which happen for his fellows. The systems of
reference appear to us to have practically everything in common and we are able consequently
to have intercourse with one another, to refer to
the actual events as common. A philosophical
problem of fundamental importance is involved
in this fact of intercourse, but it need not interrupt
our attempt to form a world-view. Whether the
world is the condition of intercourse or intercourse the condition of the existence of the
world, we think of the world as common to
all.

Our world then, or to be precise let each of us
say, my world, appears as a certain range of
activity and a certain possibility of experience.
This world is definitely limited, its limits have
been pushed out by our advancing science, but it
is on one side limited by our concept of the atomic
system, on the other by our concept of the stellar

system. Within these limits there are infinite systems of reference, using the word infinite here in its ordinary discursive meaning. For example, there is the animal world, the insect world, the vegetable world, the protozoan world, and the microbian world, each containing within it count-less different ranges of activity and innumerable possibilities of experience. Now imagine, in the manner of folk-tales, that we have the power to pass from our own system into any of these, that is, imagine that our mind can enter the living organism of bird or insect or microbe, possess its range of activity and enjoy its experience, still continuing by memory our former experience. We have then only to reflect on the nature of life and consciousness to see that the change in the system of reference cannot be a change in the subject of experience, and can only be a change in the object of experience. The subject passing to the new system of reference must therefore necessarily bring with it its norm, the standard of its measurement in reference to which it judges objects to be great or small. It is obvious that a creature, small in size as judged by me, and to whom I am a Brobdingnagian, does not feel its smallness, it feels my greatness, for its norm is supplied by its own instrument of activity, its living body. There are in fact within our ordinary perspective myriads of subjects of experience, each of which finds in its bodily organization the norm of its dimensions. If then I pass from one system to another it is certain that my space and time units must vary, for, unless they do, there is no conceivable way of effecting the exchange of standpoint. Now we have admitted that there are limits to our universe. We are bounded, we have said, by the atomic system on one side, the stellar system on the other. Even the minutest microbe is far removed from the atomic limit, and the mythical beings whom the poets have created are well within the stellar limit. Let us, however, boldly imagine that the change of system carries us right to the limit. It is easily conceivable. We have to imagine our proportions reduced to the ten thousand trillionth and we are within the atom, or, increased to the same amount and the earth is as far below the limit of perception as an electron. So that in the first case the electron of the atom has become for us a planet which will appear to us,—at least we can imagine it,—as a universe precisely like the present, and its limits will be as now, an atomic system on the one hand, a stellar system on the other. And in the other case the present stellar system will have become an atomic system. This is the way in which the infinity of the universe presents itself to me when space and time are recognized as variable and not constant.

I shall be challenged however. It will be said that I have not escaped from the dilemma of the old mathematical infinity, because though I may carry my norm of measurement as the inseparable adjunct of consciousness and vary my space and time through infinite change of system, it is after all only for me that the standard is constant, to an independent external observer I become larger or smaller absolutely. But it is precisely this idea that there can be an independent observer in an absolute system of reference that the principle of relativity negatives and rejects.

It seems to me, therefore, that the principle of relativity is a philosophical principle which is not only called for by the need of mathematical and physical science for greater precision in the new field of electro-magnetic theory in which it is continually advancing, but is destined to give us a new world-view. It will be found, as it has always been found, that the poets with their mythical interpretations, and the philosophers with their speculative hypotheses, have led the way in this new advance. The continuity of the universe can only be a continuity of consciousness, and the mode of this continuity is imaginatively presented to us in the old eastern myth of the transmigration of the soul and (may we not also say?) in the Christian mystery of the Incarnation.

I conclude, then, that in every reflection on our actual experience we are directly conscious of an objectivity which we distinguish from our subjective activity of knowing. Whether we approach the problem of that objectivity from the abstract standpoint of physical science or from the concrete standpoint of philosophy, the result is the same. Ultimately, in spite of its claim to independence, all that an object or event is, in substance or in form, it derives from the activity of the life or mind for which alone it possesses the meaning which makes it an object or event. This is not a mystical doctrine, nor is it esoteric. If we adopt in mathematics and physics the principle of relativity (and have we any choice?) the obstinate, resistant form of the objectivity of the physical world dissolves to thin air and disappears. Space and time, its rigid framework, sink to shadows. Concrete four-dimensional space-time becomes a system of world-lines, infinitely deformable. And these world-lines, do not they at last bring us in sight of an irreducible minimum of self-subsistent objectivity ? No. The world-lines are not things-in-themselves, they are only an expression for what is or may become common to different observers in the relations between their stand-points. Carried to its logical conclusion the principle of relativity leaves us without the image or the concept of a pure objectivity. The ultimate reality of the universe, as philosophy apprehends it, is the activity which is manifested in life and mind, and the objectivity of the universe is not a dead core serving as the substratum of this activity, but the perception-actions of infinite individual creative centres in mutual relation.

A closing illustration will perhaps serve better than argument to bring home to the reader the philosophical meaning of the principle. On a frosty morning we may see the watery vapour in the air we breathe condense into a small cloud and then rapidly disappear, reabsorbed into the atmosphere. Imagine that at such a moment we should undergo a sudden transformation of all our proportions so that our new dimensions become infinitesimal in comparison with our present state. Would it appear to us that we ourselves had changed ? The principle of relativity declares that the change could not possibly be experienced by us as change in ourselves because with the alteration in proportions the ratios remain constant. The change would express itself in the new dimensions of objects. The little globules of water which composed the cloud would now appear as stars and planets at immense distances from one another, undergoing a slow age-long evolution and obeying the law of the inverse square. The change would be a new space and a new time.

INDEX

Aristotle, 24, 27, 29, 51

Berkeley's *Theory of Vision*, 18

Bergson on Zeno's Problem, 34, 36

Bradley, F. H., on Zeno's Problem, 31

Clarke, The Correspondence with Leibniz, 92, 114

Conduit, Madame, The Story of Newton and the Apple, 80

Contraction Theory of Fitzgerald and Lorentz, 129

Copernican Discovery, 11, 60 ff.

Cratylus, the Disciple of Heracleitus, 29

Dante, 15, 58, 59

Darwinian theory compared with the Copernican discovery, 75

Democritus, 40 ff., 58

Descartes, 61 ff.

Dispute concerning the discovery of the Calculus, 104

Eclipse expedition of 1919, 10, 136

Einstein, 1, 4, 38, 133 ff.

Epicurus, 43, 52

Epicureans, Dante's description, 59

Equivalence, 10, 141

Euclid's Postulates, 144

Fitzgerald, Professor G. F., 129

Galileo, 149, 150

Genetic Theories of Space-perception, 19

Heracleitus, 29

Hume, 17

Hypotheses non fingo, 94, 119

James, William, 20

Kant, 2, 3; The Antinomies of Reason, 32, 33

Kepler, 82

Laplace, 101

Leibniz, 99 ff.; Correspondence with Clarke, 92; satirized by Voltaire, 101, 114 ff.

164